Geoinformatics Applications in Agriculture

Geoinformatics Applications in Agriculture

Dr. Uma Bhatnagar

Oxford Book Company
Jaipur ▪ New Delhi

Oxford Book Company

Regd. Off. "Bony Residency", Gate No. 2
Opp. Tilak Public School, Vishweswariya Nagar,
Gopalpura Road, Jaipur - 302018, India
Ph: 0141 - 2761280, Telefax: 2761381
email - oxfordbookjpr@gmail.com
website - oxfordbookcompany.com

Delhi Office: 102, 1st Floor, Satyam House,
4327/3, Ansari Road, Darya Ganj,
New Delhi - 110002, India
Ph: 011-45652440

Edition, 2025

ISBN : 9789380179629

Price : ₹ 2250.00

Typeset by Shivangi Computers, Jaipur

Printed and bound in India by Thomson Press Ltd.,
New Delhi

Preface

Agriculture resources are among the most important renewable, dynamic natural resources. Comprehensive, reliable and timely information on agricultural resources is very much necessary for a country like India whose mainstay of the economy is agriculture. With increasing population pressure throughout the nation and the concomitant need for increased agricultural production there is a definite need for improved management of the nation agricultural resources. In order to accomplish this, it is first necessary to obtain reliable data on not only the types, but also the quality, quantity and location of these resources. Geoinformatics has been and it will continue to, a very important factor in the improvement of the present systems of acquiring and generating agricultural data.

Remote sensing is nothing but a means to get the reliable information about an object without being in physical contact with the object. It is on the observation of an object by a device separated from it by some distance utilising the characteristics response of different objects to emissions in the electromagnetic energy is measured in a number of spectral bands for the purpose of identification of the object. Remote sensing techniques have a unique capability of recording data in visible as well as invisible part of electromagnetic spectrum. Therefore certain phenomenon, which cannot be seen by human eye, can be observed through remote sensing techniques i.e. the trees, which are affected by disease, or insect attack can be detected by remote sensing techniques much before human eyes see them. The increased efficiency and profitability that the proper application of technology can provide has made precision agriculture the hottest developing area within traditional agriculture.

An unique resource on geoinformatics applications in agriculture, this book examines how this powerful technology can help increase crop production and lower costs. It describes the nature of the geoinformation technologies, examines the purpose of the GIS application, and describes methods used to develop the applications. The book will be a toolkit for the acquisition, management, and analysis of spatial data throughout the agriculture value chain.

Dr. Uma Bhatnagar

Contents

1

Mapping Biophysical Factors

GIS maps and databases for global environmental conditions and agricultural productivity potential have been available for quite some time, it is only in the past few years that it has become possible to overlay these maps with georeferenced population data. This mask differentiates urban and rural areas; based on population distribution data generated by the LandScan Global Population Database (ORNL Online) and Nighttime Lights of the World (NOAA Online). It has been used to create a gridded rural area and population database at a resolution of 5 arc-minutes. All rural population numbers reported here have been derived from this database. Current suitability of land for agricultural activity depends on three factors: (i) environmental conditions, (ii) biophysical requirements of plants and animals to survive and thrive, and (iii) level and nature of human inputs. Climate, i.e., temperature and rainfall, determines whether or not growing conditions needed for specific crops are present. It is thus the first variable considered in the evaluation of biological yield potential of an area for individual crops.

POPULATION AND HUMAN SETTLEMENTS

Distinguishing Urban And Rural Area

The procedure involved the creation of an urban area mask based on two sources: (i) a GIS map showing the location of nighttime lights and (ii) the LandScan 2002 ambient population map, adjusted to UN population data for 2000. The LandScan map shows the distribution of the world's population at a resolution of 30 arcseconds (one square kilometre at the equator). The urban mask has been generated by capturing the denselypopulated gridcells, up to a threshold corresponding approximately to UN urban population figures by country for the year 2000. Gridcells not captured by the urban area mask are considered rural, even though a few of them may contain rural settlements and other built-up area. To distinguish non-settled from settled rural area, FAO created another grid which classifies rural pixels with population density greater than 2 000 persons per square kilometre as rural settlements. This threshold was selected based on an IIASA analysis that showed that almost negligible land area would be left for agriculture in areas with population density greater than this number. Due to unreliability of data for states with an area of less than 3 000 square kilometres or population of less than 500 000 and non-availability of data for Antarctica and states and territories for which the UN does not publish urban and rural population data, pixels for these areas were coded as "not assessed". Three 30 arc-second databases contain the urban area, rural area and rural settlement grids that were compiled using the method summarised above. The databases are available on the DVDs that contain the FGGD Digital Atlas for the year 2000.

Estimating the Distribution of Rural Population

By overlaying the area grids on the LandScan population distribution grid, FAO has generated three population distribution grids—for urban, rural and rural settlement populations respectively. Applications using these three databases can estimate urban, total rural, settled rural and non-settled rural populations for any area of interest.

Size of Excluded Area and Population

As the reference map and GIS software used for the GAEZ study and its refinements differ from those used by FAO/SDRN to generate its rural and urban area and population grids, there are differences in the estimation of global area and built-up area between the two sources. Another distinction has been made within total rural area, between that which is currently available for rainfed agriculture, irrespective of its suitability, and that which is not.

ENVIRONMENTAL CONDITIONS AND CONSTRAINTS

A major factor determining the vulnerability of rural populations to hunger is the quality and availability of land and water resources for agricultural production. Information about this factor has been derived from the FAO/IIASA global agro-ecological zones (GAEZ) assessment. The worldwide land resources database compiled for the GAEZ study enables an evaluation of biophysical limitations and production potentials for major food, feed, fodder and fibre crops in different environments and under various levels of inputs and management conditions. The land resources database, organised by grid cell and aggregated to national, regional, and global coverage, provides the basis for several applications. These include the following:

1. Identification of areas with specific climate, soil, and terrain constraints to crop production;
2. Quantification of potential rainfed crop yield and production under the assumptions of three levels of farming technology and management;
3. Estimation of the extent of land area suitable for rainfed and irrigated cultivation and pasture;

4. Estimation of the potential for production increase, either through bringing additional land under cultivation or through increasing input levels.

The global distribution and severity of climate, soil, and terrain slope constraints derived from the GAEZ land resources database are described briefly below, and the distribution of the world's rural population, as affected by each type of constraint, is presented.

Climate

For the GAEZ study, historical climate database layers relevant to crop and pasture production potential were created from two datasets compiled by the Climate Research Unit (CRU) at the University of East Anglia: one consisting of average climate data for the period from 1961 to 1990 and the other consisting of similar data for individual years for the period from 1901 to 1996. These data have then been used to determine the thermal climate, length of growing period and degree of climate variability for each grid cell.

Thermal Climates

The classification system for thermal climate zones used in the GAEZ study includes the following latitudinal belts: tropics, subtropics with summer rainfall, subtropics with winter rainfall, and temperate, boreal and polar/arctic belts. The temperate and boreal belts have been further subdivided according to continentality into three classes, namely: oceanic, sub-continental and continental. Table 1 gives the land area and rural population of the thermal climate zones of the world.

Table 1: Rural area and rural population of the world, by thermal climate zone

Thermal climate zone	Rural area		Rural population		Rural population density
	km² (1,000)	share of total rural area %	persons (1,000)	share of total rural population %	persons/km²
Tropics	48,613	38.5	1,397,910	43.6	29
Subtropics (summer rainfall)	17,668	14.0	899,920	28.0	51
Subtropics (winter rainfall)	12,583	10.0	185,638	5.8	15
Temperate (oceanic)	2,119	1.7	49,713	1.6	23
Temperate (sub-continental)	17,798	14.0	594,445	18.5	33
Temperate (continental)	9,152	7.2	72,365	2.3	8
Boreal (oceanic)	224	0.2	393	0.0	2
Boreal (sub-continental)	3,768	3.0	3,623	0.1	1
Boreal (continental)	11,887	9.4	2,467	0.1	0
Arctic	2,562	2.0	79	0.0	0
Total	126,374	100.0	3,206,553	100.0	25

Length of Growing Period Zones

A general characterisation of moisture conditions is achieved in AEZ through the concept of length of growing period (LGP), i.e. the period during the year when both moisture availability and temperature are conducive to crop growth. To capture the temperature component alone, the expression $LGP_{t=5}$ (temperature growing period) is used to indicate the number of days with mean daily temperature above 5 °C, i.e. conducive for crop growth. The expression LGP is then used to refer to the number of days within $LGP_{t=5}$ when moisture conditions are considered adequate. The growing period for most crops continues beyond the rainy season and, to a greater or lesser extent, crops mature on moisture stored in the soil profile. However, the amount of soil moisture stored in the soil profile, and available to a crop, varies, e.g., with depth of the soil profile, the soil physical characteristics, and the rooting pattern of the crop. The relevant values for individual soil units in a gridcell were used to set limits to available soil moisture, enabling calculation of possible extension of the growing period beyond the end of the rainy season for individual soils.

On the basis of annual climate data for the period 1901-96, the actual length of growing period (LGP) in each gridcell was calculated for each individual year, and standard deviation of LGP (SD in days) and coefficients of variation (CV in percent) were determined. Climate constraints are classified according to length of periods with cold temperatures and moisture limitations. Temperature constraints are related to the length of the temperature growing period ($LGP_{t=5}$). An $LGP_{t=5}$ of less than 120 days is considered a severe constraint, while an $LGP_{t=5}$ of less than 180 days is considered as posing moderate constraints to crop production. Within areas where $LGP_{t=5}$ does not pose a severe constraint, hyper-arid and arid moisture regimes (LGP<60 days) are considered to be severe constraints, while dry semiarid moisture regimes (LGP 60–119 days) and per-humid regimes (LGP>365 days) are considered to be moderate constraints.

Soils

Soil Types and Qualities

The FAO/UNESCO Digital Soil Map of the World provides a classification of soils into 106 soil units. These units are defined in terms of measurable and observable properties of the soil itself, many of which are directly relevant to agricultural production potential. The gridded database includes information about the percentage occurrence of soil units in each pixel; other information about the properties of each soil unit is kept in a linked soil association composition database. In order to represent them in map form, the soil units have been collapsed into 26 major soil groups.

Soil Constraints

Soil depth and soil quality, as defined below, are the attributes that are relevant for evaluation of the suitability of land for agriculture. The shallower the soil, and the poorer its quality, the more severe are the constraints to crop agriculture and grazing. Criteria used in the GAEZ study for establishing the severity of soil constraints are described below:

1. *Depth:* soils with a depth of less than 50 cm are severely constrained; those with a depth of 50–100 cm are moderately constrained. Deeper soils are not constrained.
2. *Soil chemical status and natural fertility:* soils with high salinity, sodicity or gypsum contents are severely constrained, as are soils with low natural fertility; those with moderate natural fertility are moderately constrained; those with high natural fertility are not constrained.
3. *Drainage:* soils that are poorly or imperfectly drained (gleysols, planosols, soils with antraquic phases) are severely constrained; soils with gleyic and stagnogleyic subgroups are moderately constrained; excessively and well-drained soils are not constrained.
4. *Texture:* coarse textured soils and soils with stones, boulders or rock outcrops in the surface layer or at the surface are severely constrained; soils with heavy cracking clays are moderately constrained; other soils with medium and fine textures are not constrained.
5. *Miscellaneus land:* this land is not fit for agriculture and includes: dunes, shifting sands, salt flats, rock debris, desert detritus, and glaciers and snow caps. Miscellaneous land units are classified as severely constrained.

Terrain Slope

Terrain Slope Classes

Sloping terrain is more difficult to cultivate than flatland, and is subject to higher rates of water runoff and soil erosion. Generally speaking, the steeper the slope, the greater the constraint to productivity potential, although this constraint can be relieved to some extent through use of terraces. In the context of suitability of a land area for agriculture, the unit of measurement used to define terrain slope classes is the percent slope. This refers to the rise in elevation in meters over a range of 100 meters. The Digital Soil Map of the World distinguishes three broad slope classes, namely, level to undulating (dominant slopes ranging between 0 and 8 percent), rolling to hilly (dominant slopes ranging between 8 and 30 percent), and steeply dissected to mountainous (dominant slopes more than 30 percent). As with soil types, terrain slopes also exhibit a high degree of variation within small areas.

For the GAEZ study seven slope classes were defined, and the distribution of slope classes was determined for each 30 arc-second gridcell of the Global Digital Elevation Model (GTOPO30). The results were aggregated into 5 minute gridcells and, based on known relationships between soil types and slope, into individual soil association units. From this a derived slope distribution was developed.

Terrain Slope Constraints

The slope thresholds applied in the GAEZ study to define the degree of constraint to productivity of different categories of agricultural land are as shown below. Table 2 shows the distribution of global rural area and rural population by degree of terrain slope constraint.

1. *For rainfed land:* land with slopes greater than 30 percent is severely constrained; land with slopes from 16–30 percent is moderately constrained; land with slopes 8–16 percent is slightly constrained; land with slopes 0–8 percent is not constrained.
2. *For non-terraced land with gravity irrigation:* land with slopes greater than 8 percent is severely constrained; land with slopes 5–8 percent is moderately constrained; land with slopes 2–5 percent is slightly constrained; land with slopes 0–2 percent is not constrained.
3. *For land with sprinkler irrigation:* land with slopes greater than 16 percent is severely constrained; land with slopes 8–16 percent is moderately constrained; land with slopes 5–8 percent is slightly constrained; land with slopes 0–5 percent is not constrained.

Table 2: Rural area and rural population of the world, by degree of terrain slope constraint

Terrain slope constraints	Rural area		Rural population		Rural population density
	km^2 (1,000)	share of total rural area %	persons (1,000)	share of total rural population %	persons/km^2
No constraints	16,945	13.4	444,422	13.9	26
Very few constraints	26,246	20.8	804,980	25.1	31
Some constraints	36,333	28.7	884,514	27.6	24
Some severe constraints	22,704	18.0	561,862	17.5	25
Frequent severe constraints	14,303	11.3	342,953	10.7	24
Very frequent severe constraints	8,557	6.8	152,130	4.7	18
Unsuitable for agriculture	1,286	1.0	15,692	0.5	12
Total	126,374	100.0	3,206,553	100.0	25

Environmental Constraints To Rainfed Agriculture

Global Occurrence of Combined Constraints

The GAEZ land resources database allows characterisation of various regions according to the prevailing climate, soil and terrain constraints. Based on this, a constraint classification has been formulated and has been applied to each gridcell of the database. On the basis of currently available soil, terrain and climatic data, the GAEZ study estimated that some 10.5 billion hectares of land, i.e., almost four-fifths of the global land surface (excluding Antarctica) suffer rather severe constraints for rainfed crop cultivation.

Distribution of Rural Population by Constraint Class

The presence of constraints reduces the productivity potential of the land below the maximum yield potential of the crop or crops being grown. However, unless the constraints are severe, production and sustainable agriculture is possible in the presence of constraints, and in fact, this is the condition in which most agriculture is practiced around the world. Except in locations that are absolutely too cold or too dry for rainfed agriculture, even those with severe terrain and soil constraints can sustain some agriculture, at least for a few years until the soils are exhausted or eroded.

LAND PRODUCTIVITY POTENTIALS

Procedure For Estimating Productivity Potential of Land for Rainfed Crops and Pasture

The characterisation of climate, soil and terrain conditions relevant to agricultural production provides a first indication of the degree of suitability of land areas for rainfed agriculture, based on geophysical factors. However, this covers only one face of the total picture. The techniques of suitability analysis have been developed to permit consideration of the more complex interplay of geophysical, biological and socioeconomic factors that actually determine productivity potentials in real life conditions.

Definition of Land Utilisation Types (LUTs)

The GAEZ method uses crop models and environmental matching procedures to identify the degree of crop-specific environmental limitations across the world's land area, under three different assumptions about the input levels and management conditions. The crop models use the environmental requirements and adaptability characteristics of each crop species assessed and calculate the expected yields under different sets of environmental conditions and input/management scenarios, while the input/management scenarios reflect different investment options that may be implemented by humans. These are combined by specifying a number of crop production systems, termed land utilisation types, or LUTs. A LUT consists of a set of technical specifications within a socioeconomic setting. As a minimum requirement, both the nature of the produce and the setting must be specified. Attributes specific to particular land utilisation types include crop information such as cultivation practices, input requirements, crop calendars, and utilisation of main produce, crop residues, and by-products. For the GEAZ studies completed thus far, 171 crop LUTs have been defined, each with defined environmental requirements and adaptability characteristics for a specific crop type. Table 3 shows the crops and crop types included in the nine crop groups. Global productivity potential is assessed for each crop type, at three different levels of inputs and management under rainfed conditions, and two different levels of inputs and management under irrigated conditions.

Table 3: Crop Groups And Crops Crop Types Thermal Climate Zones

Crop Groups and Crops	Crop Types	Thermal Climate Zones
Cereals	(83)	
Wheat (hibernating)	4	Boreal, Temperate and Subtropics
Wheat (non-hibernating)	12	Boreal, Temperate, Subtropics and Tropics
Rice, japonica (wetland)	4	Tropics, Subtropics and Temperate
Rice, indica (wetland)	4	Tropics and Subtropics
Rice (dryland)	3	Tropics
Maize (grain)	13	Tropics, Subtropics and Temperate
Maize (silage)	6	Subtropics and Temperate
Barley (hibernating)	4	Boreal, Temperate and Subtropics
Barley (non-hibernating)	12	Boreal, Temperate, Subtropics and Tropics
Sorghum	7	Tropics, Subtropics and Temperate
Pearl millet	2	Tropics
Foxtail millet (setaria)	4	Subtropics and Temperate
Rye (hibernating)	4	Temperate and Subtropics
Rye (non-hibernating)	4	Boreal, Temperate and Subtropics
Fibre crops	(7)	
Cotton	7	Tropics, Subtropics and Temperate
Oil Crops	(25)	
Soybean	6	Tropics, Subtropics and Temperate
Rape (hibernating)	2	Temperate and Subtropics
Rape (non-hibernating)	6	Temperate, Subtropics and Tropics
Groundnut	3	Tropics, Subtropics and Temperate
Sunflower	6	Temperate, Subtropics and Tropics
Oil palm	1	Tropics and Subtropics
Olive	1	Subtropics and Temperate

Countd.

Countd.

Pulses	(17)	
Phaseolus bean	9	Tropics, Subtropics and Temperate
Chickpea	5	Subtropics and Tropics
Cowpea	3	Tropics
Roots and Tubers	(14)	
White potato	4	Boreal, Temperate, Subtropics and Tropics
Cassava	1	Tropics
Sweet potato	3	Subtropics and Tropics
White yam	2	Tropics
Greater yam	2	Tropics
Yellow yam	1	Tropics
Cocoyam (taro)	1	Tropics
Stimulants	(7)	
Cocoa	1	Tropics
Arabica coffee	1	Tropics
Robusta coffee	1	Tropics
Tobacco	4	Tropics, Subtropics and Temperate
Sugar Crops	(6)	
Sugarcane	1	Tropics and Subtropics
Sugar beet	5	Temperate and Subtropics
Fruit Crops	(1)	
Banana/Plantain	1	Tropics and Subtropics
Citrus	1	Tropics and Subtropics
Vegetables	(10)	
Cabbage	2	Tropics, Subtropics and Temperate
Onion	4	Tropics, Subtropics and Temperate
Tomato	4	Tropics, Subtropics and Temperate
Total	171	

Estimation of Maximum Attainable and Long-term Achievable Crop Yields

Ranges for maximum attainable yields ranges were calculated for each crop present in tropical, sub-tropical and temperate/boreal zones. The maximum attainable yields for each crop under rainfed conditions represent averages of simulated year-by-year yields attainable during the period 1960 to 1996. With balanced fertiliser applications and proper pest and disease management (which is best possible at high level of inputs), only limited fallow will be required to maintain soil fertility and to keep pest and disease outbreaks in check. At low level of inputs, assuming no application of chemical fertiliser and only limited used of organic material, and very limited or no application of biocides, there is need for considerable fallow periods in the crop rotations to restore soil nutrient status and to break pest and disease cycles. In the GAEZ study, long-term achievable yields are calculated by applying a fallow-period requirement factor dependent on climatic conditions, soil type, crop, and level of inputs/management. The yields attained in the long-term are well below the estimated maximum attainable yields when accounting for fallow period requirements.

Estimation of Land Productivity Potential for Rainfed Cereals

Table 4 compares maximum (short-term) attainable yields and long-term achievable yields on suitable land for staple cereals (rainfed wheat, rice, grain maize). On average, long-term achievable yields are 10, 20, and 55 percent lower than maximum attainable yields, respectively at high, intermediate and low levels of inputs. These represent the best estimate of the productivity potential of land for this crop group.

Table 4.2: Maximum (short-term) attainable and long-term achievable yields

Major area	Low inputs		Intermediate inputs		High inputs	
	Short-term attainable t/ha	Long-term achievable t/ha	Short-term attainable t/ha	Long-term achievable t/ha	Short-term attainable t/ha	Long-term achievable t/ha
Africa	1.0	0.4	3.7	3.0	6.4	5.8
America						
Latin America and Caribbean	1.2	0.6	3.7	3.1	5.8	5.3
North America	0.8	0.4	3.6	2.8	5.8	5.2
Asia (*including Japan*)	1.0	0.5	3.5	2.8	5.3	4.8
Europe						
Europe (excluding Russian Federation)	1.0	0.4	4.0	3.1	6.4	5.8
Russian Federation	0.7	0.3	2.9	2.5	4.4	4
Oceania (*including Australia and New Zealand*)	0.7	0.4	3.2	2.6	5.3	4.8
Developed	0.9	0.4	3.5	2.8	5.6	5.1
Developing	1.1	0.5	3.7	3.0	6.2	5.6
World	1.0	0.4	3.7	3.0	5.9	5.4

Variability of Rainfed Cereal Production

The historical climate data for individual years have been used to calculate a time series of production potentials for cereals. Of a total of 83 cereal types, consisting of cultivars of wheat, rice, barley, rye, sorghum, millet, and setaria, AEZ tests and selects for each of the 2.2 million gridcells of the land resources database for each year the cereal type that results in the highest production, given the climatic conditions for that year. During the 20th century, the global average of potential cereal production, based on year-by-year simulations for currently cultivated land with fixed inputs, increased by over five percent between the period 1901-30 and the period 1961-90. On the other hand, simulations based on average climate data for the whole period from 1961 to 1990 produced a figure for average potential cereal production for the globe that was almost 10 percent higher than the average obtained using historical climate data for individual years. This comparison provides a rough estimate of the losses being incurred due to climate variability.

Estimation of Crop and Pasture Suitability Indices

For each of the approximately 2.2 million gridcells of the database suitability results were estimated for each crop type, at each of the three input levels considered under rainfed conditions and for gridcells classified as irrigated area for each of the two input levels considered under irrigated conditions. The outcomes were then mapped by means of a suitability index (SI). This index reflects the suitability make-up of a particular gridcell. In this index VS represents the portion of the gridcell with attainable yields that are 80 percent or more of the maximum potential yield for the specified input scenario. Similarly, S, MS and mS represent portions of the gridcell with attainable yields 60–80 percent, 40–60 percent, and 20–40 percent of the maximum potential yield, respectively. SI is calculated using the following equation:

$$SI = VS*0.9 + S*0.7 + MS*0.5 + mS*0.3$$

where: VS = very suitable; S = suitable; MS = moderately suitable; mS = marginally suitable

The resulting database permits generation of a suitability map based on the crop suitability index (CSI) for each gridcell for any combination of crop types that the analyst specifies.

To estimate production potential of the global land surface for pasture, estimates of maximum potential and attainable pasture yields were generated and a pasture suitability index (PSI) at low input level under rainfed conditions was obtained for each gridcell in the land resources database, using the same AEZ matching procedures that were employed for estimating rainfed crop production potential. For the GAEZ study, a reassessment of global pasture production potential was made by combining the AEZ matching procedure results and the zonal features of grassland composition. First, the original AEZ biomass and yield calculation procedures were revised to better cope with, in particular, arid and hyper-arid environments. This modification entailed the incorporation of monthly rainfall events, available from CRU climate database. This enabled more realistic assessments of moisture regimes in space and time. Another modification involved the replacement of the LGP-based biomass and yield estimation in arid zones by an enhanced Net Primary Productivity (NPP) calculation procedure, which was applied in the arid environments prevailing in zones with LGP of less than 30 days. These two modifications have significantly improved correspondence with satellite-derived data for arid pasture and shrub areas.

Suitability of Currently Available Land for Rainfed Crops and Pasture

Estimation of Land Area Currently Available for Rainfed Agriculture

The fact that an area of land has been determined to be suitable for rainfed crop agriculture or pasture does not necessarily mean it is available for such use. A sizeable part of the more suitable area is covered by forests, and is likely to remain so for the foreseeable future. Some suitable land, much of it forested, has been set aside as protected area, for environmental or other reasons. Also, although the extents involved are small, in some locations human settlements have been built on land that would otherwise be suitable or very suitable for agriculture. Finally irrigated area, although representing highly productive agricultural land which may also be used for rainfed crops as well, has been kept separate from land currently available for rainfed agriculture. Although suitability analysis has also been done for irrigated area, only suitability under rainfed conditions, as these are the conditions where agriculture is likely to be more vulnerable and poor and hungry people are likely to be found. Hence, land area currently available for rainfed agriculture excludes area belonging to the above-mentioned land cover categories.

Suitability Results

Rainfed suitability has been calculated for nine crop groups that are important to most farming systems in developing countries, namely, cereals, fibre crops, oil crops, pulses, roots and tubers, stimulants, sugar crops, tree fruits and vegetables. The algorithm examines in each gridcell all the crop types belonging to a particular crop group. Among these it determines the crop type that maximises agronomic suitability.

Multiple Cropping Zones

In the AEZ crop suitability analysis, the LUTs considered refer to single cropping of individual crop types, i.e., each crop type is presumed to occupy the land only once a year and in pure stand. In areas where the growing periods are sufficiently long to allow more than one crop to be grown in the same year or season, single crop yields do not reflect the full potential of total time and space available per unit area of land for rainfed production.

To assess multiple cropping potential, a number of multiple cropping zones have been defined through matching both growth cycle and temperature requirements of individual suitable crop types with time available for crop growth. For rainfed conditions this period is approximated by the LGP, i.e., the number of days during which both temperature and moisture conditions permit crop growth. Table 5 presents the occurrence of these multiple cropping zones on currently available land in major world regions.

Table 5: Occurrence of multiple cropping zones under rainfed conditions, currently available land

Major area and region	Currently available land km² (1,000)	Zone A	Zone B	Zone C	Zone D	Zone E	Zone F	Zone G	Zone H
		%		share of currently available land					
AFRICA									
Eastern Africa	5,595	16.6	50.1	15.8	14.0	–	1.6	–	1.9
Middle Africa	4,060	18.2	23.0	7.2	42.5	–	7.3	–	1.8
Northern Africa	7,704	76.2	18.2	3.3	2.3	–	–	–	–
Southern Africa	2,360	50.6	41.2	5.5	1.4	0.7	0.3	0.3	–
Western Africa	5,685	46.4	28.4	6.9	14.7	–	3.3	–	0.3
AMERICAS									
Caribbean	156	–	–	3.2	62.2	–	21.8	–	12.8
Central America	2,230	27.9	34.7	7.4	18.4	–	7.8	–	3.8
Northern America	16,196	51.7	34.9	3.7	3.3	2.0	1.2	2.8	0.4
South America	9,992	21.5	20.8	5.4	32.6	1.5	5.2	5.2	7.8
ASIA									
Eastern Asia	9,402	52.8	24.5	3.6	3.3	5.2	4.9	4.5	1.2
Japan	88	–	47.8	12.5	19.3	17.0	1.1	2.3	–
South-central Asia	8,812	45.7	46.4	3.2	2.9	1.0	0.4	0.2	0.2
South-eastern Asia	3,158	0.1	0.6	1.4	48.1	2.3	6.7	0.2	40.6
Western Asia	4,152	73.6	26.1	0.2	0.1	–	–	–	–
EUROPA									
Eastern Europe	11,173	58.5	40.9	0.6	–	–	–	–	–
Northern Europe	1,005	42.7	57.3	–	–	–	–	–	–
Southern Europe	1,002	1.9	87.6	9.5	1.0	–	–	–	–
Western Europe	655	7.0	91.0	1.8	0.2	–	–	–	–
OCEANIA									
Australia and New Zealand	7,332	63.1	34.3	1.1	0.8	0.5	0.1	0.1	–
Melanesia	273	–	–	–	8.1	–	7.3	–	84.6
Developed	37,451	53.5	39.6	2.3	1.7	1.0	0.5	1.2	0.2
Developing	63,579	41.2	28.4	5.3	14.8	1.3	3.2	1.5	4.3
World	**101,030**	**45.7**	**32.6**	**4.2**	**9.9**	**1.2**	**2.2**	**1.4**	**2.8**

Productivity Potential of Currently Available Land for Rainfed Crops and Pasture

Productivity Potential for Rainfed Crops with Maximising Technology Mix

When considering all crop types modelled in AEZ and applying a maximising technology mix, results show that a little less than one-quarter of the Earth's land surface can be regarded as suitable for rainfed crop cultivation. When suitable land that is not currently available is deducted, Table 6 shows that only 22.7 percent of the Earth's land surface is both suitable and available, 36 percent of it in developed regions and 64 percent in developing regions. The share of currently available land that has medium high to very high potential for rainfed crop cultivation is by far the highest in Western Europe and Caribbean where the figure is respectively 66 and 51 percent. The lowest shares occur in Western Asia, Australia and New Zealand and Southern Africa where the figures fall below 10 percent. These differences only become significant, however, if scarcity of good agricultural land forces people onto less suitable and therefore less productive land, where obtaining a sustainable livelihood from agriculture may be more difficult, if not impossible.

Table 6: Extents of currently available land

Major area and region	Currently available land	Available land with medium high to very high suitability for rainfed cultivation		Available land with very low to medium suitability for rainfed cultivation	
	km² (1,000)	km² (1,000)	share of currently available land in the region %	km² (1,000)	share of currently available land in the region %
AFRICA					
Eastern Africa	5,595	2,227	39.8	2,803	50.1
Middle Africa	4,060	1,866	46.0	1,438	35.4
Northern Africa	7,704	808	10.5	2,044	26.5
Southern Africa	2,360	164	6.9	1,054	44.7
Western Africa	5,685	1,328	23.4	1,612	28.4
AMERICAS					
Caribbean	156	80	51.3	75	48.1
Central America	2,230	337	15.1	1,351	60.6
Northern America	16,196	3,855	23.8	5,237	32.3
South America	9,992	4,104	41.1	3,684	36.9
ASIA					
Eastern Asia	9,402	1,134	12.1	3,086	32.8
Japan	88	32	36.4	56	63.6
South-central Asia	8,812	1,206	13.7	4,537	51.5
South-eastern Asia	3,158	1,064	33.7	2,071	65.6
Western Asia	4,152	239	5.8	1,297	31.2
EUROPA					
Eastern Europe	11,173	2,824	25.3	1,655	14.8
Northern Europe	1,005	286	28.5	274	27.3
Southern Europe	1,002	369	36.8	596	59.5
Western Europe	655	433	66.1	175	26.7

Countd.

Countd.

OCEANIA					
Australia and New Zealand	7,332	493	6.7	2,671	36.4
Melanesia	273	85	31.1	181	66.3
Developed	37,451	8,292	22.1	10,664	28.5
Developing	63,579	14,642	23.0	25,233	39.7
World	101,030	22,934	22.7	35,897	35.5

Productivity Potential of Currently Available Land for Rainfed Pasture

Recent estimates of pasture and grazing areas available globally suggest that almost a quarter of the earth's land surface is covered by pasture or shrub vegetation that could, and often does, provide feed resources for ruminants. Results of the pasture suitability assessments for currently available land are given in Table 7. Worldwide, about 19 percent of available land has medium high to very high potential for pasture – slightly less than the aggregate amount with medium high to very high potential for rainfed crops. However, the share of available land area that has at least marginal suitability for pasture is double the share for rainfed crops –around 70 percent for pasture compared to 35 percent for crops.

Table 7: Extents of currently available land with potential for pasture, by region

Major area and region	Currently available land km² (1,000)	Available land with medium high to very high suitability for pasture km² (1,000)	share of currently available land in the region %	Available land with very low to medium suitability for pasture km² (1,000)	share of currently available land in the region %
Africa					
Eastern Africa	5,595	1,493	26.7	4,071	72.8
Middle Africa	4,060	1,205	29.7	2,470	60.8
Northern Africa	7,704	89	1.2	3,855	50.0
Southern Africa	2,360	142	6.0	2,206	93.5
Western Africa	5,685	492	8.7	3,582	63.0
Americas					
Caribbean	156	144	92.3	12	7.7
Central America	2,230	496	22.2	1,733	77.7
Northern America	16,196	3,339	20.6	11,477	70.8
South America	9,992	3,865	38.7	5,793	58.0
Asia					
Eastern Asia	9,402	1,895	20.2	6,418	67.4
Japan	88	76	86.4	11	12.5

Countd.

Countd.

South-central Asia	8,812	374	4.2	7,305	85.3
South-eastern Asia	3,158	1,910	60.5	1,242	39.5
Western Asia	4,152	47	1.1	3,534	85.1
Europa					
Eastern Europe	11,173	1,915	17.1	8,906	79.7
Northern Europe	1,005	312	31.0	693	69.0
Southern Europe	1,002	262	26.1	735	73.3
Western Europe	655	485	74.0	166	25.3
Oceania					
Australia and New Zealand	7,332	413	5.6	6,852	93.5
Melanesia	273	198	72.5	75	27.5
Developed	37,451	6,802	18.2	28,840	77.0
Developing	63,579	12,350	19.4	42,296	66.7
World	101,030	19,152	19.0	71,136	70.5

Distribution of Rural Population on Currently Available Agricultural Land

The difference in the distribution of good to prime land across regions reflects mainly the distribution of cold barren land, desert and tropical rainforest among the continents. What is more important is the pattern of human settlement. In developed countries, 26.4 percent of the rural population live in areas with prime agricultural land, and only 11.3 percent live in marginal areas. In developing countries, by contrast, only 10.5 percent live in areas with prime agricultural land, whereas 23.3 percent live on land that is only marginally suitable for rainfed agriculture, even with an intermediate level of inputs and management. This result suggests that, even with substantial investment to raise inputs and management from their current low level to an intermediate level, the prospects for alleviating poverty through sustainable agricultural growth in these marginal areas is slim. And if investment in a high level of inputs and management are made, this will almost certainly mean introducing large-scale commercial agriculture, with current small–scale farmers becoming paid farm workers or leaving agriculture.

REFERENCES

Climatic Research Unit (CRU). 2002. "A high-resolution data set of surface climate over global land areas", by M.G. New, D. Lister, M. Hulme & I. Makin. *Climate Research*, 21: 1-25.

EROS Data Centre (EDC). 2000. *Global land cover characterisation (GLCC)*. Version 2.0 (available at http://edcwww.cr.usgs.gov).

IIASA. 2002. *"Climate change and agricultural vulnerability"*, by G. Fischer, M. Shah & H. van Velthuisen. Laxenburg, Austria.

Zhang, X. & Zhou, G. 1995. "A new NPP model", in Ye, Duzhong, Lin, Hei, et al. (editors). *China Contribution to Global Studies: China Global Change Report* 2. Beijing: Science Press.

2

Precision Farming

The *Global Positioning System* (GPS) is the heart of precision agriculture. A GPS receiver is a location device that calculates its position on earth from radio signals broadcast by satellites orbiting the earth. The U.S. government has 24 satellites that are constantly orbiting the earth. These satellites contain precise atomic clocks, and the exact time is encoded into the signals broadcast from each satellite. A GPS receiver uses this time information to measure the distance to each satellite from which a signal is being received. With at least four satellite signals, the receiver can use triangulation to calculate its position on the ground. However, this calculated position is inaccurate because of errors in the satellite signals. Some of these errors, such as those caused by atmospheric interference, are unavoidable. Another error source is government controlled for security purposes. This intentional degradation of the satellite signals is known as selective availability. Figure 1 shows a sprayer equipped with a 60-foot boom receiving positioning data with inaccuracies of up to 300 feet from its true position. These inaccuracies are not acceptable for site specific management. Errors of this magnitude require that the government-based system be augmented with another signal that can improve positioning accuracy. The additional signal is transmitted from a fixed base station with a precisely known position. The base station receives the satellite signals and compares its calculated position with its exact position. The amount of error is then broadcast to mobile receivers in the field so that they can correct for the same satellite errors. This system is known as differentially corrected GPS or DGPS. The accuracy of the DGPS receivers is 5 to 10 feet, which is acceptable when dealing with a 60-foot swath. (Fig. 2).

Mapping software is used to handle, display and analyze data stored as a value and a position. Mapping software is available with a wide range of capabilities. Low-end packages are used primarily for creating maps or graphical images, and have little capability to process or analyze data. High-end products are known as *Geographic Information Systems* or GIS, and have many data processing capabilities. Because precision farming requires a relatively high level of data processing, software used for this purpose has become known generically as GIS software.

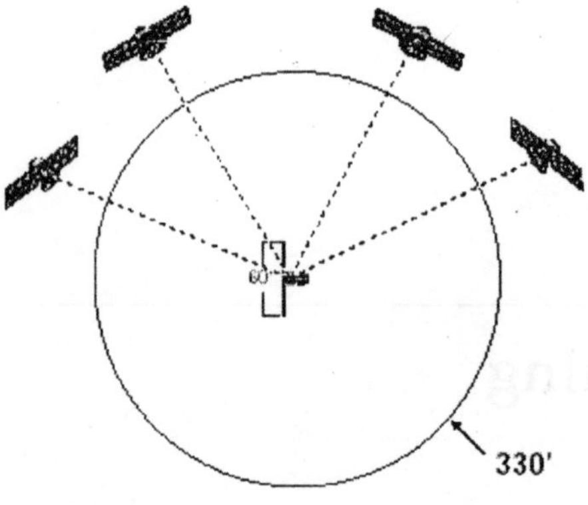

Figure 1. Position uncertainty for an uncorrected GPS receiver relative to machine operating width.

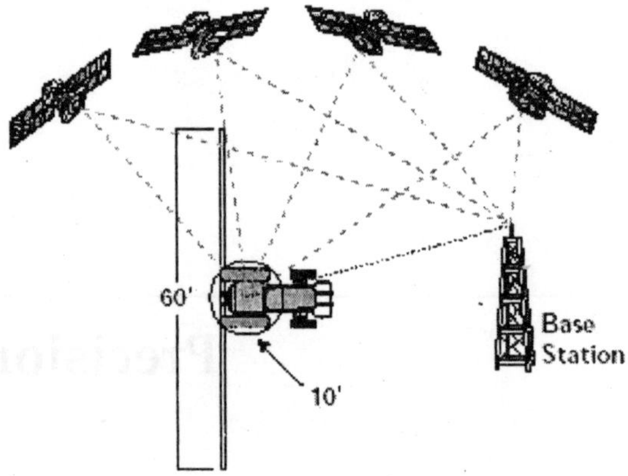

Figure 2

Potential users should consider capabilities such as data editing, interpolation/ contouring functions and statistical analysis when purchasing mapping software. Several types of data can describe a field's characteristics. Common data types include yield, soil texture and nutrient status. Each of these is stored as a data layer. Conceptually, a field can be described by a stack of data layers. Some layers may be raw data, representing the actual measurements taken (e.g., soil pH as determined by sampling and analysis or weed densities as determined by field scouting). Other layers may be derived by mathematically processing one or more layers to generate a new layer.

Variable Rate Technologies (VRT) describes machines that can automatically change their application rates in response to their position. VRT systems are available for applying a variety of substances including granular and liquid fertilisers, pesticides, seed and irrigation water. The most widely recognised VRT machines are large chemical applicators that control up to 11 different materials at once. VRT applicators consist of a controller that adjusts the actual material flow rate, a positioning system, and a map of the desired application rates for the field (Fig. 3). The controllers are very similar to those used on many sprayers, spreaders and other agricultural machines. On conventional machines, the operator controls the application rate by selecting the desired rate from the console panel in the cab. By integrating GPS and GIS databases into the system, application rate changes can be made automatically as the vehicle crosses the field.

Often the data used to determine the application rates for VRT machines is gathered through grid sampling of fields. *Grid sampling* involves taking soil samples from the field in a fixed pattern as shown in Figure 4. Each grid square typically has an area of 21/ 2 to 4 acres for grain crops, but more samples can be taken for higher value crops. Some producers have had success with *targeted* or *zone sampling*. This method uses knowledge of field conditions to determine the sampling sites rather than a uniform spacing. Generally, the samples are located where soil or topology changes occur in the field.

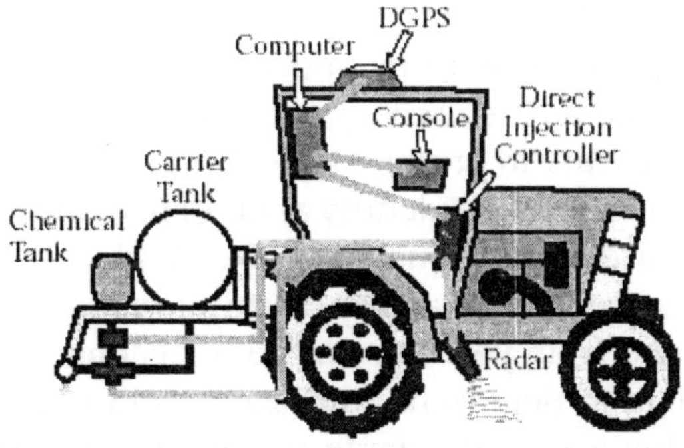

Figure 3: Schematic representation of the components of a VRT chemical applicator.

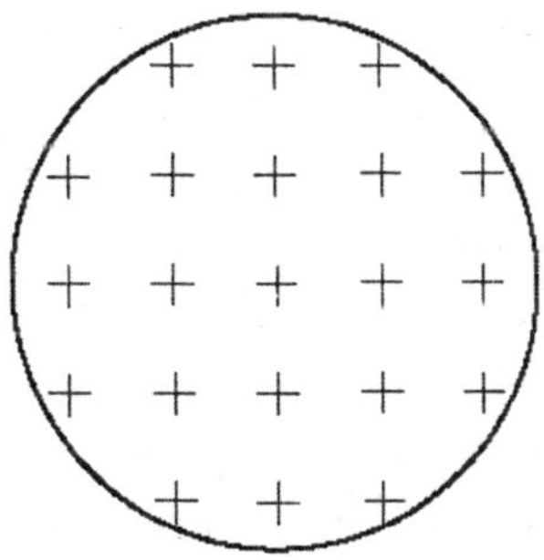

Figure 4. Each cross represents a soil sample position in a field.

The sampling data are entered into a GIS database and a map of the field's soil parameters is made. Because data is available for only a few points in the field, the values for other locations are estimated based on the values of the nearby sample points. This technique is most frequently used to determine fertiliser application rates. The required nutrient levels are determined, based on expected yield. Then nutrients are analyzed for each sample point and credit is given for these residual nutrients in each part of the field. The overall profitability of this approach to fertilisation varies by region and crop, but experiences in Texas and other states have shown that, for granular fertiliser applications on high yielding corn, costs can be reduced by $5 to 15 per acre.

Yield mapping is another important technique in precision farming. Yield maps show the variability in yield within a field. A yield mapping system measures and records the amount of grain being harvested at any point in the field, along with the position of the harvester. To produce a yield map, the harvester must be equipped with a GPS receiver and a yield monitor. A yield monitor can be a flow meter or a scale. In harvesting most grains, a sensor is placed in the flow of the grain as it passes through the harvester. This type of system measures the flow rate and moisture content of the grain, shifts the data in time to match the position where the grain was cut and divides the flow rate by the machine's harvested area to achieve a yield per unit of area (e.g., bushels per acre). For non-grain crops such as potatoes or sugar beets, scales are used to continuously weigh the produce to achieve the same results. The yield data are sent to the onboard computer where measured yield is matched with its appropriate field position and the data are stored in a memory card. The data stored on this card can be transfered to a computer equipped with mapping software to produce a yield map. Yield maps can provide useful information to the producer. The maps identify areas of high and low yields so that inputs can be adjusted to maximise the productivity of a field. Yield maps document both natural and man-made sources of variability. Natural variability is caused by weather within a growing season and from year to year. To correct for this, a producer may need to obtain several year's data to determine consistent yield trends that can be related to soils or field topology. However, man-made variability may be easily identified and corrected with a single year's data. Examples of man-made variability include poor distribution of irrigation water and the effects of past production practices.

The way harvesting machinery is operated can influence the accuracy of the resulting yield maps. For example, in irregularly shaped fields, the harvester may end up harvesting grain over a partial width of the machine. This will cause the yield sensors to indicate a lower yield than was actually obtained. Operators must understand how yield monitors record yield data. Most combine yield mapping systems are designed so that data recording is halted when the header is lifted. This makes it easy for the operator to stop recording data when traveling. However, if turns are made with the header in the down position, data points are added that show a zero yield. Another manmade error occurs when the yield sensors have not been properly calibrated. Poor sensor calibration can lead to inaccurate yield calculations. The profitability of precision farming is as variable as field conditions. In highly uniform fields, better knowledge of soil and plant parameters is not as likely to result in greater economic return as it is in fields with variable conditions. Producers who use site-specific management must recognise that information becomes another input to the system, and that it has a cost. With soil, weed, fertility and yield maps for a particular field, the producer can know more about the field's yield potential and determine which areas of a field are creating the largest profit, as well as which areas are not capable of producing as well as others. The profitability of adding more fertiliser or decreasing a pesticide rate at a specific site in the field can be determined. Precision farming technology brings increased efficiency to crop production. By closely matching application rates with crop needs, profit potential can be increased and possible environmental impacts can be minimised.

PRECISION AGRICULTURE: GLOBAL POSITIONING SYSTEM (GPS)

GPS Technology

Global Positioning System (GPS) receivers provide a method for determining location anywhere on the earth. Accurate, automated position tracking with GPS receivers allows farmers and agricultural service providers to automatically record data and apply variable rates of inputs to smaller areas within larger fields. A GPS receiver can be compared with a simple AM or FM radio. A GPS receiver "listens" for the signals that are broadcast from the satellites of the United States Department of Defense (DOD) Global Positioning System. Orbiting around the earth at an altitude of 12,550 miles, these satellites are in predictable locations; hence, we refer to the system of satellites as the GPS constellation.

Each satellite broadcasts almanac information containing the position of all satellites in the constellation. GPS receivers use the almanac to determine the position of the satellites. Minor variations in the orbits of the satellites occur due to gravitational forces from the sun and the moon. The DOD continuously monitors the satellites and adjusts the almanac information to represent the actual orbits of the satellites. The broadcast signals also contain a precisely timed predictable code that a GPS receiver can use to determine how long the signal required to reach the receiver. A microprocessor within a GPS receiver uses these delays and the position of the satellite to calculate the distance to each satellite, and then uses this information to determine location through triangulation. Triangulation is a mathematical method for locating points on a plane in three-dimensional space. If the distances to each of three satellites and your approximate location on the earth are known, the GPS receiver can calculate its terrestrial position. If information from four satellites is available, elevation can also be determined.

Accuracy

The accuracy obtained generally depends on five factors:

1. proper installation,
2. the degree of technology used in the receiver,
3. the number and location of satellites,
4. errors introduced by selective availability (SA), atmospheric conditions, the troposphere, the ionosphere, and multipathing—radio signals bouncing off objects in the area, and
5. differential corrections.

Installation

GPS antennas should be mounted on the centerline of a combine, tractor or truck and above any part of the machinery that might obstruct a line of sight to a satellite. If the cab is centered and the top of the cab is above other portions of the machine, a cab-top mounting may be the best location. However, on a steep side slope, a high mounting point will result in an error in position calculation due to the offset in horizontal position. GPS and DGPS receivers may have separate antennas but usually there will be a combination antenna so that both are centered at the same location. A delay of several seconds often occurs in agricultural applications such as yield monitoring, spraying and fertiliser application.

Example: If the antenna on a sprayer traveling 10 mph is mounted 30 ft ahead of the booms and a rate change at the controller is effected at the boom two seconds later, the rate change will occur when the booms reach the location of the antenna where the change was made. At any other ground speed, the rate change at the booms will not occur at the same location as the controller. A time adjustment must usually be factored into the system to compensate for time delays in sensing or product application.

Electrical interference can result from electrical storms, power lines, 2-way radios, nearby radio transmitters, electric motors, microwave towers, cellular phones, vehicular electrical equipment such as alternators and ignition systems on spark-ignition engines, and other sources. Changing the position of the antenna or adding noise suppression kits may reduce interference problems from alternators and ignition systems. Follow the instructions for installation of the GPS equipment, making sure that all connections are tight.

Technology

Low-cost receivers receive signals from one satellite at a time and require more time to determine the location than a receiver capable of receiving four signals simultaneously. Usually, seven to 10 satellites are in view at any one time and more sophisticated receivers will produce the most accurate location. *Reacquisition time* is the time it takes to get an accurate position fix after a short-time loss of satellite signals; this may occur for a variety of reasons, including traveling near trees or buildings and losing the 'line of sight' to satellites. Reacquisition time is important for most agricultural applications and, especially, for guidance with applicators and aircraft. New technology in GPS receivers has shortened reacquisition time. Receivers that can track 8-12 satellites are less susceptible to acquisition loss.

Satellite Constellations

Using triangulation to calculate position, small errors in distance can cause large errors in position. The error in calculating position through triangulation increases when the satellites are close together. The best accuracy is produced when the receiver can pick up signals from many widely dispersed satellites (Figure 5).

Selective Availability and Other Errors

To prevent an enemy from using GPS satellite signalsfor determining locations on earth, the DOD "scrambles" the signals sufficiently to introduce an error of about 100 meters in an uncorrected location calculation. The term for this is "selective availability" (SA).

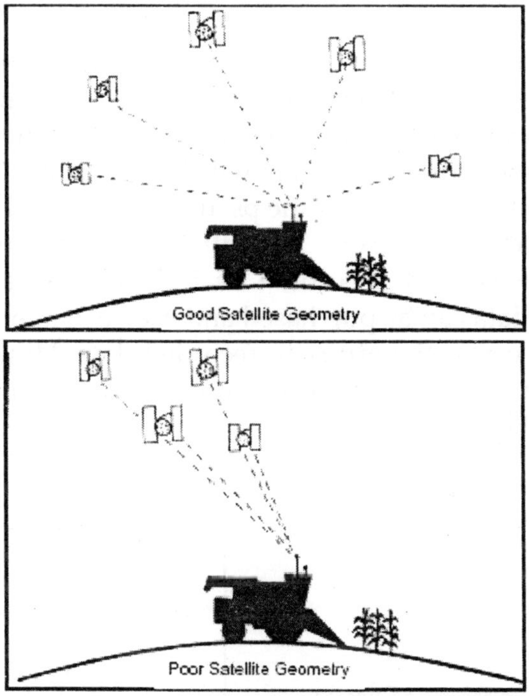

Figure 5. Good satellite geometry yields more accurate GPS position estimates.

Atmospheric, tropospheric and ionospheric conditions, however, also cause distortions or errors in calculating distance; natural errors due to these conditions are not easily or reliably predicted. Hence, even in the absence of SA, differential corrections will still be required to accurately calculate position. Multipathing, the phenomenon that creates distorted television signals is caused by signals that bounce off of other objects before reaching the antenna (Figure 6). Multipathing cannot be corrected by differential corrections.

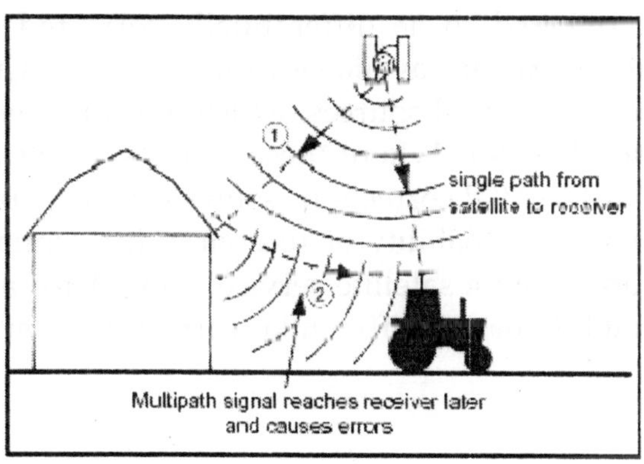

Figure 6. GPS signals can somtimes "bounce" off of objects.

Differential Corrections

Stationary GPS receivers are used to calculate the total error due to SA, variable atmospheric conditions and other factors. The concept is simple. A stationary receiver always has a known location; because the actual positions of the satellite and the receiver are known, the true range (distance) is known. The distance calculated by the receiver using the broadcast signals is known as the pseudorange, which is generally in error due to the combined sources of all errors. The difference between the true range and the pseudorange is the error and is known as the differential correction (Figure 7). Differential correction data can be purchased and used at a later time in a process known as post processing to correct the errors in recorded data. However, the most common approach is to connect a differential corrections receiver to a GPS receiver to provide real-time corrections (Figure 8).

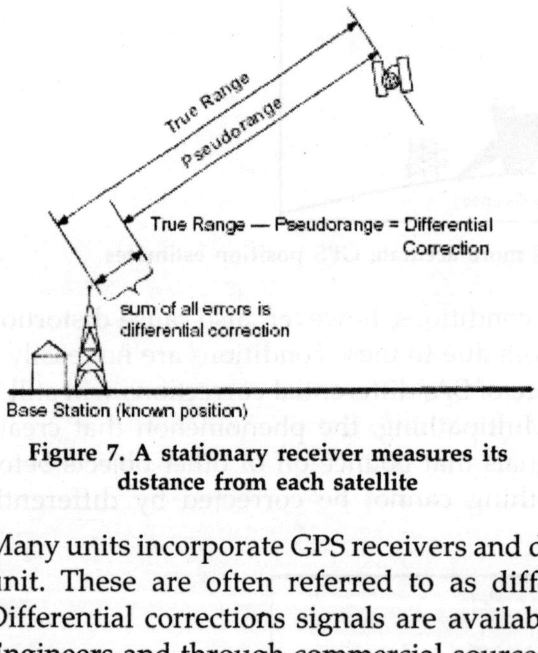

Figure 7. A stationary receiver measures its distance from each satellite

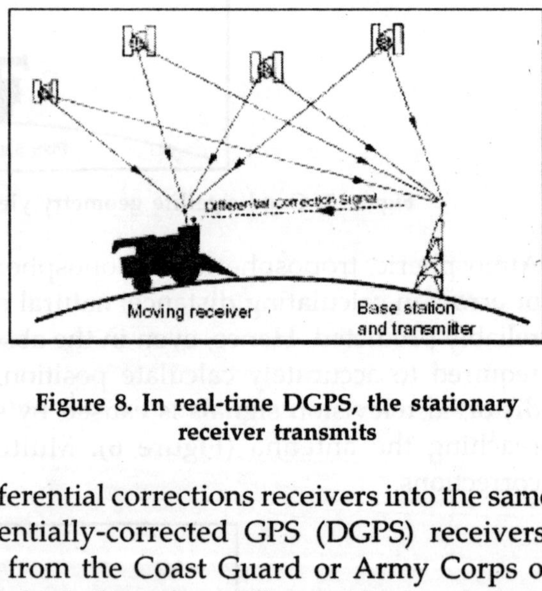

Figure 8. In real-time DGPS, the stationary receiver transmits

Many units incorporate GPS receivers and differential corrections receivers into the same unit. These are often referred to as differentially-corrected GPS (DGPS) receivers. Differential corrections signals are available from the Coast Guard or Army Corps of Engineers and through commercial sources, which, for a fee, will provide signals from a satellite or a land-based tower. Where these sources aren't available, or for special applications, a private differential corrections source can be installed. Some of the newer DGPS receivers combine the capability of receiving differential signals from both the Coast Guard beacons and from a satellite service. Table 1 for a comparison of features of Coast Guard and satellite-based differential corrections sources.

Table 1. Comparison of Coast Guard and satellite differential correction sources by feature.

Feature	Coast Guard beacon	Satellite Differential
Accuracy (RMS)	<1 m (depends on distance from area beacon station)	<0.75 m uniform over service - depends on service provider
Initial equipment cost	lower initial cost	higher initial cost
Annual subscription cost	none in USA but signal not available in many regions of the US	$500 to $1000 per year, depending on level of service
Interference susceptibility	subject to local man-made noise sources	minimal interference from man- made sources
Range	100 to 250 miles	large coverage area- most of US

Coast Guard signals

The Coast Guard signals are broadcast in the frequency range of 285-325 kHz (just below the usual AM-radio band) where radio waves travel as ground waves and are not limited to line-of-sight reception like FM-radio stations. The signals are series of pulses similar to those of the GPS satellites. Referred to as Minimum Shift Keying modulation, the signal is less sensitive to electrical interference and noise than AM-radios.

Missouri has free access to correction signals from Coast Guard beacons located near St. Louis (@ 322 KHz), Kansas City (@ 305 KHz), Tulsa (@ 299 KHz), Rock Island (@ 311 KHz), Memphis (@ 310 KHz) and Omaha (@298 KHz).

The range of the Coast Guard beacons is approximately 150 miles in good weather (electrical storms cause interference). Accuracy decreases with distance from the transmitter. This service is expected to become the choice of many agricultural users, especially in Missouri where several signals are available. A disadvantage of the Coast Guard differential corrections signal is the rate at which the beacon transmits or repeats messages. Most Coast Guard sites broadcast at 200 bits per second. At this broadcast rate, the age of a satellite's differential correction can be as old as four seconds. For some applications, such as guidance, this update rate may be unacceptable. For guidance applications, update rates of two to ten times per second may be required. Typical Coast Guard beacon receivers have two channels. One channel receives the differential correction and the other is searching for the best incoming signal. This helps to ensure against loss of a DGPS signal if at least two beacons are within range.

Satellite-based Correction Signals

For the user, one of the simplest types of differential corrections signals is transmitted from a geostationary satellite. Companies such as Omnistar, Accqpoint and Racal provide this service. The typical annual user's fee ranges from $500 to $800. The correction signal is available throughout most of North America. The accuracy of high-quality receivers is generally considered to range from one to three meters RMS. Interference from man-made sources is minimal. Satellite-based signals may have an advantage for operation around trees and buildings since the satellite is nearly overhead at most locations and within the DGPS receiver's line-of-sight.

Land-based Correction Signals

Several commercial land-based correction signal services are also available for a fee. Some companies put up their own transmitters to broadcast correction signals; these include SatLoc, Mobile Data and CSI. Some commercial service providers piggyback correction signals onto commercial FM radio station transmitters. These sub-carriers include Pinpoint Communications, DCI and others.

Private GPS Receiver and Radio Transmitter

GPS users not covered by Coast Guard or commercial sources of differential corrections can install a stationary receiver and transmitter to provide their own differential corrections source. Few users in Missouri will choose to buy and install their own fixed GPS receiver and transmitter since the Midwest has other choices available.

Cost vs. Accuracy

The accuracy attainable with GPS depends partly on how much you are willing to spend, ranging from approximately $100 to $100,000. A low-cost (from $100 to $500) GPS receiver without DGPS capability may be sufficiently accurate for some crop scouting applications, for navigating highways or for locating your favorite fishing spot on a lake. The RMS horizontal accuracy may be about 50 yards.

The cost for a basic DGPS receiver suitable for most agricultural applications is about $3,000 to $5,000 and provides RMS accuracy of at least three meters with a typical accuracy of one meter, which is sufficient for yield monitoring and grid soil sampling. If you need a GPS receiver for guidance (for spraying, fertiliser application, etc.), the cost may be up to $25,000. Such systems provide accuracy down to a few inches. Since sprayers and fertiliser spreaders can travel fairly quickly, lower quality GPS equipment may not update position quickly enough to be used for guidance or control, although GPS systems with high update rates and accuracies in the range of one foot or less are becoming available at lower prices. The annual subscription cost for some differential correction services varies with the level of service (accuracy). Some providers offer three levels of service, e.g., one provider has a premium service for better than 1 meter accuracy, intermediate service for accuracies in the range of 5 meters and a basic service for accuracies in the range of 10 meters. Typical approximate costs may be $600, $250 and $75 per year, respectively, depending on the level of service.

Coordinate Systems

Several coordinate systems are in use for mapping and may cause problems with compatibility between software systems. Users frequently need to transform position data into a plane (flat) coordinate system, either to merge them with another data set, to plot a map of the GPS results, or to perform further calculations for such parameters as area, distance or direction (plane coordinate systems are usually easier to work with than geodetic coordinates). When using data and maps from several sources, coordinates must be based on the same datum. The coordinate system differences, which are caused by a different reference frame, ellipsoid and data adjustment, are significant (up to several hundred meters) and cannot be ignored. Several commercially-available software programmes produced by well-known GIS vendors treat the coordinate shifts incorrectly. The National Geodetic Survey provides software (LEFTI and NADCON) at a nominal charge to compute datum shifts.

GPS receivers can usually report position information in more than one format. The most common format is lat/lon (latitude and longitude). Lat/lon coordinates are recorded in angular units of degrees, minutes and seconds. One second of latitude is equal to about 30 meters. GPS receivers may display lat/lon in degrees plus minutes to four decimal places (instead of minutes and seconds). Most geographic information system (GIS) software is capable of using more than one format and may automatically convert lat/lon coordinates to a coordinate system such as Universal Transverse Mercator (UTM) or State Plane Coordinates (SPC) to calculate distances in meters or feet.

UTM and SPC systems project portions of the earth's curved surface onto a flat map and report locations as actual distances from a reference point in meters and feet, respectively. Hence, no conversions are necessary to calculate distance or area. Commercial software available from several GPS vendors will compute UTM or state plane coordinates from GPS data. These coordinates are usually based on the WGS-84 datum and thus are in the NAD-83 system. If these must be transformed to NAD-27, it is advisable to do the NAD 83 to NAD 27 transformation in geodetic coordinates, and then make the conversion to plane coordinates as the final step.

Universal Transverse Mercator Coordinates

The UTM coordinate system is a worldwide system originally adopted by the U.S. military in 1947, and since has been widely used by civilian mapping in many countries. The UTM system is consistent throughout the world and one set of equations will allow calculation of coordinates at any location. The world is divided into 60 zones each spanning 6 degrees in longitude and extending north and south from a latitude of south 84 degrees to north 84 degrees.

State plane coordinate system

In the 1930s, the U.S. Coast and Geodetic Survey established a plane coordinate system for each of the 48 states. One to five zones were established in each state with a Lambert Conformal or a Traverse Mercator projection.

REFERENCES

Differential GPS Explained. 1996. Trimble Navigation Ltd., P.O. Box 3642, Sunnyvale, CA 94088-3642.

GPS: A Guide to the Next Utility. 1993. Trimble Navigation Ltd., P.O. Box 3642, Sunnyvale, CA 94088-3642.

The Precision-Farming Guide for Agriculturalists. 1997. John Deere Publishing , Dept. 374, John Deere Road, Moline, IL 61265-8098. Phone 1-800-522-7448.

The State of Site-Specific Management or Agriculture. 1997. Published by: the American Society of Agronomy, Inc.; the Crop Science Society of American, Inc.; and the Soil Science Society of American, Inc.

Target Farming: A Practical Guide to Precision Farming Concepts and Technology. 1996. Written and published by Ron Johnson, A.Sc.T., 511 Haslam Cres., Saskatoon, Sask S7S 1E7 Canada, Phone & Fax: 306-665-1610.

3

GIS Solutions for Farming

All sectors of the agriculture industry use ESRI's geographic information system (GIS) technology to share data, increase yields, predict outcomes, and improve business practices. By applying GIS technology to their operations, agricultural operations are able to manage resources and responsibilities more efficiently, devise data portals that disseminate vast amounts of agricultural data and interactive maps, and support farming communities. Producers use GIS to better manage their farms by creating information-dense reports and maps that give them a unique perspective of their operations. The powerful analytical capabilities of GIS offer an array of options for visualising farming conditions, as well as measuring and monitoring the effects of farm management practices. Combined with remote-sensing technology, GIS can be used to precisely determine and control inputs, saving preventive expense and reducing the amount of harm to the soil. Farm managers also use GIS to submit government programme applications, simplifying what used to be time-consuming multistep processes. ESRI's GIS technologies support people working in agriculture by providing :

1. Greater analytical support for precision farming
2. Better understanding of risk factors
3. Higher revenue generation and cost recovery
4. Greater efficiency through task automation
5. Greater access to government services and data
6. More accurate support for decision making
7. Greater insight to policy making
8. Easier reporting for government applications and regulatory compliance
9. Better resource management

GIS FOR PRECISE FARM MANAGEMENT

Monitoring market trends, improving yields, and predicting weather are among the many responsibilities required to reduce the risk of loss and increase profitability. The *Farmer's Almanac* has been replaced with geospatial analysis and predictive modeling. With these tools at their disposal, farmers now have the ability to visualise their land, crops, and management practices in unprecedented ways for precise management of their businesses. Today, accessing spatial data has become an essential farm practice. Government agencies such as the U.S. Department of Agriculture (USDA) and the European Union host Web sites that deliver valuable information to help farmers better understand their land and make more informed decisions. This data can be accessed on the Internet and used to create intelligent maps for better farm business practices.

GIS FOR GOVERNMENT AGRICULTURE

Federal, state, and local agricultural agencies serve and share information with many different groups. Throughout the supply chain, from producer to consumer, agricultural ministries and institutions collect, collate, manage, and disseminate data. With growing public awareness of farming practices and legislation, these activities will inevitably become more complicated. If there is a common theme in new efforts to better conduct these activities, it is communication. A powerful medium of communication, GIS enables the sharing of information not only internally but also between government and consumer. The agricultural industry uses GIS to meet the challenges of farm management including regulatory compliance, permit distribution, subsidy tracking, and pest management. By improving the way producers collect, store, and access information related to these activities, GIS enables a more complete command of their businesses.

AGRICULTURE INFORMATION DELIVERED ONLINE

In the agriculture industry, those unacquainted with GIS technology need to view and print custom maps and be able to download these maps to desktop and GPS-enabled mobile applications. Using ESRI® GIS software, AgTerra Technologies Inc., a manufacturer of agriculture management applications located in Sheridan, Wyoming, set out to make this happen. AgTerra uses its Internet map server programme in conjunction with its AgTrac application to make it easier to build an online map that can be downloaded as an ArcGISArcExploreror ArcPad® project. To make system data more accessible to workers, AgTerra uses ArcIMS® for many reasons, the foremost being its highly scalable framework for GIS Web publishing. ArcIMS gives users the ability to serve maps and other data to a variety of clients ranging from wireless devices, such as cellular phones and personal digital assistants (PDAs), or lightweight browser-based clients to full-featured GIS desktop clients. As a result of this project, AgTerra improved AgTrac's ability to manage and facilitatc improved business.

PUBLISHING VOLUMES OF DATA

In August 2005, the USDA Natural Resources Conservation Service (NRCS) launched the Web Soil Survey, a Web-based version of the National Cooperative Soil Survey. The Web Soil Survey helps the USDA meet two of the U.S. president's management agenda items: to eliminate paperwork and to make government services accessible to more people. Prior to the Web Soil Survey going live, soil survey maps and data were largely contained in soil survey books housed at local USDA service centers, NRCS field offices, and public libraries. The Web Soil Survey application was built using ESRI ArcIMS, ArcGIS Server, and ArcSDE® software. The server software is being fully integrated with the soil survey's master database, which contains all soil information on a national basis. ArcIMS is the front end for delivering the application, while ArcGIS Server delivers the information. In addition, ESRI's ArcWebSM Services provide address-finding functionality, enabling users to type an address and receive a map view of that area.

An application suite, the Web Soil Survey contains a front-end application that accesses the Soil Data Mart—the centralised repository of the Web Soil Survey's spatial and tabular soil data. Because the Web site provides single-point access, users can access available historical and current soil survey data. Past publications are available as PDFs for viewing or downloading for selected areas. On the interactive GIS map, the user can outline an area of interest (from 3 to 10,000 acres) and select a usage option. The application provides suitability levels of the area for the selected land use. The Web Soil Survey offers approximately 50 national standard interpretations. Some examples of interpretations are land applications; productivity of cropland, rangeland, and forestland; and suitability for different recreational developments such as paths and trails, campsites, and picnic grounds. Adding new functionalities to the Web Soil Survey project is a constant endeavor. Eventually, USDA customers will be able to log in to their USDA case files and drop in information, such as property boundaries, for quick access to information about selected properties in their files. The Web Soil Survey is the beginning of a new era for NRCS and delivery of soil survey information.

IMPROVING CROP YIELD

Fertiliser Application System Uses GIS/GPS for Precision Farming

Today more than ever, farmers rely on inputs to increase the quality of the produce they grow. These inputs vary from seed, fertiliser, and pesticides to machinery parts and monitoring equipment. In past generations, farmers tended to spread fertiliser evenly over the entire field. Now, thanks to geospatial technology, spreading methods can be more exact by type, quantity, and location of application. Outfitting the spray rig with computerised controllers and GPS navigation systems is an approach to farming that is both friendly to the environment and profitable for the farmer. Ravensdown, New Zealand's largest manufacturer and distributor of fertilisers, is using geospatial technology to improve the way New Zealand farmers manage their inputs. By using GIS and GPS to guide the application of fertilisers, farmers are decreasing the amount of wasted resources that can potentially cause harmful runoff into streams and waterways. At the same time, they are reducing their total fertiliser expenditure by up to 10 percent per year.

Ravensdown worked with Eagle Technology Group to design the solution that is built on ESRI's ArcGIS software. The system accurately records where and how much fertiliser has been applied to a certain area. This information is merged with digital orthophotos and the farm's relational databases to create a vivid picture of the farm's overall soil sustainability. Ravensdown's GIS architecture includes ESRI's ArcGIS Server, ArcSDE, ArcIMS, and ArcExplorer. Once the raw spatial and attribute data is captured from GPS transceivers on the trucks, it is transmitted wirelessly to Ravensdown's facility where it is loaded into the GIS and processed in near real time. The onboard GIS creates a map-based display that shows fertiliser application data as a series of colour-coded "snail trails" that are overlaid on the map, giving a very good representation of the process. An additional benefit is that the system can be used as evidence to verify "proof of placement." With geospatial technology, it is easy to demonstrate that the fertiliser has been spread in a manner consistent with best environmental practices.

Eagle used the development capabilities available in ArcGIS Server to embed the map interface in Ravensdown's customer relationship management (CRM) system. Ravensdown wanted seamless access to the spatial and attribute data, so Eagle developers used the .NET framework to build an interactive map viewer that the call center staff can access with a click of the mouse. Staff members can query the database, manipulate the display, and print or fax hard-copy maps. Ravensdown is planning to build on this foundation. "Eventually, we want to incorporate GIS into our quoting system to calculate the road distance between our depots and the fields to be fertilised," says Mark McAtamney, CIO at Ravensdown. "The distance is a significant component of the cost to our clients."

GIS: The Geographic Approach

ESRI software offers scalable solutions for agricultural departments, universities, and businesses. From field-based products such as ArcPad to server-level software such as ArcGIS Server, data can be collected and managed more easily than ever before. ArcGIS provides all the necessary tools to analyze the spatial components of agricultural datasets. Below are a few examples where organisations are benefiting from taking a geographic approach to their operations.

Invasive Species Response—Washington State

The gypsy moth is one of the worst agricultural pests in North America today. To deal with the burgeoning threat, the Washington State Department of Agriculture (WSDA) recently implemented GIS and GPS technology into the state's eradication efforts. As a result, the department has reduced the amount of paper used in the data collection process and increased the number of moth catches, significantly improving WSDA's monitoring efforts. The accuracy of GPS units, combined with ArcView® software's state-of-the-art mapping capabilities, gives trappers a better picture of where to set the traps to better manage this voracious pest.

Vineyard Inspection—Washington State

Chateau Ste. Michelle Wine Estates maintains a wealth of spatial data for its numerous vineyards. To keep the grapes growing at peak level, crews routinely inspect the vineyards using ESRI's ArcPad mobile platform. Data on the mobile devices is synchronised using TC Technology's GO! Sync™ for ArcPad. The remotely captured data is stored centrally in an enterprise geodatabase, allowing Chateau Ste. Michelle to analyze a vast amount of information relating to the numerous conditions that affect the health of its vines such as weed growth, disease presence, and pest infestation. This field-collected data is then synchronised with the winery's centrally stored data.

Mass Animal Burial Soil Suitability—Kansas

The quick disposal of large dead animals following a catastrophe is critical. The pit method disposes dead animals by placing carcasses in successive layers in an excavated pit. The chemical and physical properties of soils can help distinguish areas that are most suitable for excavating pits and disposal of large animals as well as areas with the least number of soil limitations. Using ArcGIS Desktop software, the USDA obtained soil data from the National Soil Information System and combined it with the Soil Survey Geographic Database to map the soil suitability for large-animal disposal pits. This interpretive map demonstrates how important soils are in determining which areas may be suitable for these emergency response uses.

REFERENCES

The Precision-Farming Guide for Agriculturalists. 1997. John Deere Publishing , Dept. 374, John Deere Road, Moline, IL 61265-8098. Phone 1-800-522-7448.

The State of Site-Specific Management or Agriculture. 1997. Published by: the American Society of Agronomy, Inc.; the Crop Science Society of American, Inc.; and the Soil Science Society of American, Inc.

Target Farming: A Practical Guide to Precision Farming Concepts and Technology. 1996. Written and published by Ron Johnson, A.Sc.T., 511 Haslam Cres., Saskatoon, Sask S7S 1E7 Canada, Phone & Fax: 306-665-1610.

4

Data Sharing in Agriculture

For decades, the federal government has tried to reduce duplicative geospatial data collection by coordinating GIS activities within and outside the federal government. For example, in 1990, the Office of Management and Budget established the Federal Geographic Data Committee to promote the coordinated use, sharing, and dissemination of geospatial data nationwide. In 1994, the National Spatial Data Infrastructure (NSDI) programme was established by executive order to address the problem of the redundancy and incompatibility of geospatial information on a national basis. More recently, Geospatial One-Stop, a component of NSDI, was initiated. Although efforts to build the NSDI are progressing, achieving the vision of a nationwide GIS network remains a formidable challenge. Developing standards that meet stakeholders' needs remains a challenging and timeconsuming task, and achieving full participation across governments in their development has also been difficult.

Geospatial One-Stop is aimed at promoting coordinated geospatial data collection and maintenance across all levels of government. Among its objectives are :

1. deploying an Internet portal for one-stop access to geospatial data;
2. developing data standards; and
3. encouraging greater coordination among federal, state, and local agencies.

While these objectives are important, Geospatial One-Stop has focused on limited, near-term tasks and was not intended to fully address the longer-term challenges of implementing the NSDI. A much more substantial effort will be required to attain the broader vision of seamless integration of GIS data nationwide. Existing draft standards may need further revision, and more extensive coordination efforts may be required to ensure broad adoption at all levels of government. Further, the effort is likely to require a continuing effort over an extended period of time, due to the fact that significant investments have already been made in existing non-standard systems.

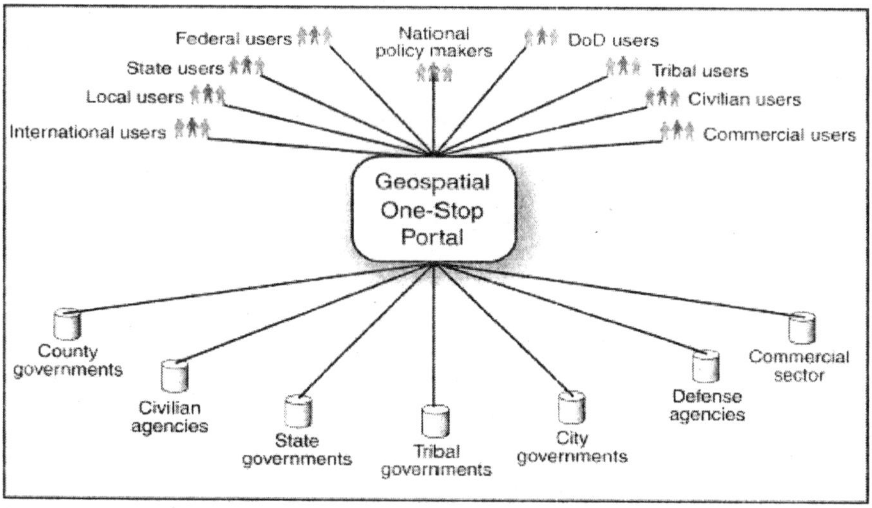

Source: GAO based on FGDC information

Figure 1: Geospatial One-Stop Portal Concept

The primary function of a GIS is to link multiple sets of geospatial data and graphically display that information as maps with potentially many different layers of information. Assuming that all the information is at the same scale and has been formatted according to the same standards, users can potentially overlay spatial information about any number of specific topics to examine how the layers interrelate. Each layer of a GIS map represents a particular "theme" or feature, and one layer could be derived from a data source completely different from the other layers. For example, one theme could represent all the streets in a specified area. Another theme could correspond to all the buildings in the same area, and others could show vegetation or water resources. As long as standard processes and formats have been arranged to facilitate integration, each of these themes could be based on data originally collected and maintained by a separate organisation. Analyzing this layered information as an integrated whole can significantly aid decision makers in considering complex choices, such as where to locate a new Department of Motor Vehicles building to best serve the greatest number of citizens. Figure 2 portrays the concept of data themes in GIS.

The expansion of Internet connectivity in recent years has substantially enhanced the potential value of GIS because now it is possible to locate and harness data from many disparate GIS databases to develop very rich analytical information on almost any topic that is associated with physical locations. Data that were once collected and used only for a single purpose could now have much broader applications. Further, the community of GIS users has been broadened to include potentially anyone with an Internet connection. For example, citizens can now use home computers to obtain answers to specific questions about land use in their state or local jurisdiction.

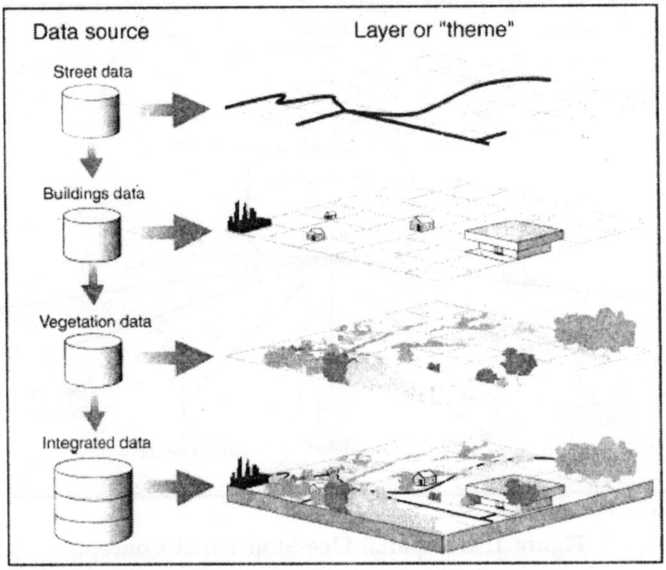

Figure 2: GIS Layers or "Themes"

Commercial entrepreneurs can combine GIS data about zoning and tax-incentive areas to determine what parts of a city are best suited for establishing a new business. Federal, state, and local government agencies are using GIS today to provide vital services to their customers. For example, local fire departments can use geographic information systems to determine the quickest and most efficient route from a firehouse to a specific location, taking into account changing traffic patterns that occur at various times of day. Highway departments use GIS to identify intersections that have had a significant number of personal injury accidents to determine needs for improved traffic signaling or signage. GIS can also be an invaluable tool in ensuring homeland security by facilitating preparedness, prevention, detection, and recovery and response to terrorist attacks.

Many federal departments and agencies use GIS technology to help carry out their primary missions. For example, the Department of Housing and Urban Development (HUD) worked with the Environmental Protection Agency (EPA) to develop E-MAPS, which combines information on HUD's community development and housing programmes with EPA's environmental data. The programme provides homeowners and prospective homebuyers with ready access to detailed local information about environmental hazards that otherwise would likely have been very difficult to obtain. In another example, the Department of Health and Human Services uses GIS technology to analyze data on population and topography (including roads, streams, and land elevation), as well as information gathered from residents. These data are used to track the spread of environmental contamination through a community, to identify geographic areas of particular health concern, and to identify susceptible populations, such as children or the elderly.

The usefulness of GIS in disaster response was demonstrated recently in connection with the Space Shuttle Columbia recovery effort. After the loss of Columbia on February 1, 2003, debris was spread over at least 41 counties in Texas and Louisiana. Analysis of GIS data was critical to the efficient recovery and documentation of that debris. The Texas state GIS programme provided authorities with precise maps and search grids to guide field reconnaissance and collection crews. Officials in charge of the effort used maps of debris fields combined with GIS data about the physical terrain to carefully track every piece of debris found. This information not only contributed to an efficient recovery effort but also may help experts to refine theories about why Columbia perished.

MANY FEDERAL GOVERNMENT GIS ACTIVITIES OVERLAP

According to the Department of the Interior, about 80 percent of all government information has a geospatial data component, such as an address or other reference to a physical location. It is not surprising, then, that a wide variety of geospatial data collection efforts are ongoing throughout the federal government, each established for a different purpose but often collecting and maintaining the same or similar information. In fact, according to the 2001 initial business case for Geospatial One-Stop, about 50 percent of the federal government's geospatial data investment is redundant. For every GIS application, federal agencies must manage the geospatial data that are at the heart of that application. In many cases, agencies maintain the same data that are referenced to the same geographic location. For example, both HUD and the Census Bureau maintain essentially the same geospatial data regarding congressional districts, city boundaries, railroads, interstate highways, and state highways. The two agencies maintain separate GIS systems for storing and analyzing this information.

In many cases, agencies independently collect data that, while not identical, is similar and potentially duplicative in many respects. For example, both the Department of Energy (DOE) and the Nuclear Regulatory Commission (NRC) collect and analyze information regarding uranium-milling facilities, and they both cover the same geographic areas. DOE concentrates on tracking the status of former uranium processing sites, whereas NRC gathers and maintains information on current uranium milling facilities in the same mid-western locations. In cases such as this, significant efficiencies may be gained by coordinating the two collection efforts. In other cases, data may be collected in different resolutions or with different degrees of accuracy but still essentially cover the same theme over the same geographic area. Local governments often possess the most recent and highest resolution geographic data; however, these data often are collected to serve specific missions and may be difficult to use for other purposes. For instance, when the Forest Service created a nationallevel GIS for the forest ecosystem, it was faced with reconciling data from a variety of incompatible locally developed systems. Local agencies had used a variety of standards for each forest and district. In assembling these data into a unified, coherent database, the Forest Service had to adopt the lowest-resolution format in order to maintain full coverage of all forests. As a result, much of the higher-resolution content of the local data could not be used. Much of the effort in building this system was spent reconciling data sets to make them usable in an integrated database.

The biggest problem with collecting this duplicative geospatial data is its cost. According to a recent study, up to 80 percent of GIS costs are related to the collection and management of geospatial data. In 1993, OMB performed a data call in which it estimated that $4.1 billion was spent annually, at the federal level, on collection and management of geographically referenced data. In addition, state and local governments are estimated to spend twice that of the federal government on collection and management of geographic referenced data.

MANY ATTEMPTS HAVE BEEN MADE TO COORDINATE GIS ACTIVITIES

The federal government has tried for years to reduce duplicative geospatial data collection by coordinating GIS activities both within and outside the federal government. In 1953 the Bureau of the Budget first issued its Circular A-16, encouraging expeditious surveying and mapping activities across all levels of government and avoidance of duplicative efforts. More recently, the E-Government Act of 2002 directed the Office of Management and Budget to coordinate the development of standard protocols for sharing geographic information to reduce redundant data collection and promote collaboration and the use of standards. Although progress has been made over this 50-year span, much work still remains to be done.

DEVELOPING AND IMPLEMENTING GIS STANDARDS HAVE POSED LONG-STANDING CHALLENGES

Developing common geospatial standards to support vital public services has proven to be a complex and time-consuming effort. The number of types of geospatial data and the complexity of those data make developing geospatial standards a daunting task. For example, 34 different broad categories of geospatial data, called "data themes," were identified in OMB Circular A-16 as a necessary foundation for the NSDI. These basic themes relate to all types of services provided by the federal government— including climate, flood hazards, federal land ownership, public health, soils, and transportation. Each of these themes, in turn, may have any number of subthemes. The transportation theme, for example, includes such divergent subthemes as road, railroad, air, transit, and waterway, each the domain of a different organisation or group of organisations. For data associated with the NSDI's themes and subthemes to be effectively shared, standards must be developed that allow interoperability and integration of the many disparate formats of data that are currently collected for each theme and subtheme. Circular A-16 further identifies seven of the themes as the core set of most commonly used data, called "framework themes." FGDC has been working to coordinate the development of these themes as well as other standards since it was established 13 years ago. Although

FGDC has developed versions of several of these standards, it has not attempted to finalise a complete set of the seven framework standards. These framework standards would define the simplest level of geographic data commonly used in most geospatial data sets.

OMB Circular A-16 calls for a well-coordinated effort among federal, state, local, and tribal governments, academic institutions, and the private sector to build an effective NSDI. Yet in the capital asset plan for the Geospatial One-Stop project published in January 2003, the Department of the Interior noted that the risk was high that agencies would be unwilling to adopt framework data standards. Given that most federal agencies—including large agencies such as DOE, Justice, and Health and Human Services—have not participated in the NSDI framework standards development process, the risk is substantial that the proposed standards will not meet their needs. In addition, agencies could be faced with a potentially expensive effort at "migrating" to the new standard. Substantial investments have already been made to independently develop geospatial systems using formats and standards that meet the specific needs of the agencies that developed them. The potential for agencies to continue to deploy agency-specific, noninteroperable geospatial systems was another high risk identified by Interior in its January 2003 Geospatial One-Stop plan.

Many states and localities have established Web sites that provide a variety of location-related information services, such as updated traffic and transportation information, land ownership and tax records, and information on housing for the elderly. Existing commercial products using a variety of formats are already meeting the needs of the states and localities in providing this information. Hence these organisations are likely to have little incentive to adopt potentially incompatible federal standards that could require substantial new investments. According to Arizona's state cartographer, many local governments currently do not comply with existing FGDC standards because most of their GIS applications were created primarily to meet their internal needs, with little concern for data sharing with federal systems.

GEOSPATIAL ONE-STOP'S OBJECTIVES ARE LIMITED

Geospatial One-Stop is intended to accelerate the development and implementation of the NSDI by promoting coordination and alignment of geospatial data collection and maintenance across all levels of government. Specifically, its objectives include: (1) deploying an Internet portal for one-stop access to geospatial data as an extension to the NSDI Clearinghouse network (see figure 3); (2) developing data standards for the seven NSDI framework data themes; (3) creating an inventory of federal data holdings related to the seven framework themes; and (4) encouraging greater coordination among federal, state, and local agencies about existing and planned geospatial data collection projects.

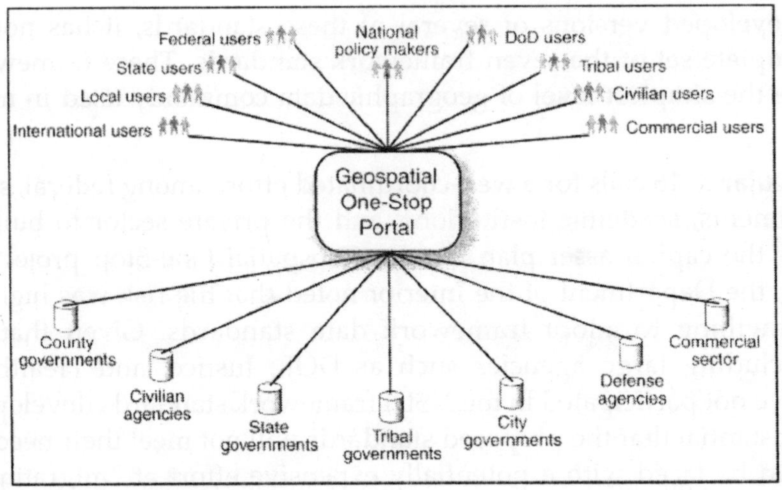

Source: GAO based on FGDC information.

Figure 3: Geospatial One-Stop Portal Concept

1. *Deploying an Internet portal*. This task was to design and implement an Internet portal to serve as a one-stop interface for users seeking links to geospatial data that were already available and cataloged in the NSDI clearinghouse. A demonstration version of the portal has been developed, and the first publicly available version is expected to be implemented by the end of June 2003. Plans are to begin adding new data to the portal, now that it has been developed. Project officials are also considering future enhancements to the functionality of the portal; however, no milestones have been set for any specific enhancements.

2. *Developing data standards*. The specific objective was to draft the seven NSDI framework standards. Drafts of these seven framework standards, as well as five transportation subthemes and a base standard have now been completed. Project officials plan to submit these drafts to the American National Standards Institute by the end of September 2003.

3. *Creating an inventory of federal data holdings*. To meet this objective, metadata for all relevant federal data sets must first be collected and made available in the NSDI Clearinghouse. Users need metadata to determine whether a data set is useful for their purposes and to be aware of any special stipulations about processing and interpreting the data. Accordingly, OMB Circular A-11 required that all federal data sets with a replacement value exceeding $1 million be documented in FGDC metadata and the metadata be accessible and searchable in the NSDI Clearinghouse network by February 10, 2003.

4. *Encouraging greater coordination among federal, state, and local agencies*. To support this objective, a process has been established to coordinate Geospatial One-Stop's activities across these various government levels. According to the project's

cooperating states coordinator, eight federal agencies are participating in developing and implementing the initiative. In addition, an intergovernmental board of directors was established with two-thirds of the vote held by state, local and tribal representatives. The purpose of the board was to help ensure collaboration among potential stakeholders from all government sectors.

According to the National States Geographic Information Council's (NSGIC) representative, state, county, and municipal levels of government are well represented and play a useful role in providing alternative views about the direction of the initiative. While Geospatial One-Stop's objectives are important, they do not represent a significantly new or different approach to the GIS integration problem that the government has been struggling with for more than a decade. First, while developing and implementing an Internet portal may offer users additional functionality over the existing Clearinghouse, unless the underlying geospatial data offered through the portal are standardised across data providers, the additional functionality offered by the portal may be of limited value. Second, the objective of finalising the seven framework standards, while important, is limited. A total of 34 data themes was identified in OMB Circular A-16 as a necessary foundation for the NSDI. Geospatial One-Stop's objectives do not include plans to address any of the remaining 27 themes. Before the broader vision of a unified nationwide network of geospatial data and systems can be achieved, standards for all of NSDI's foundation data themes will need to be established. Further, definition of the standards is only the first step in gaining their benefits;

Third, creating a complete and useful inventory of federal data holdings will require much more substantial work than is planned through the Geospatial One-Stop initiative. For example, according to the FGDC Metadata Coordinator, the extent to which agencies have posted metadata about their geospatial data sets is unknown. In addition, obtaining complete metadata from all federal sources is likely to be very challenging. If the metadata were not created when the data were originally captured, they could be expensive and time-consuming to generate after the fact, and agencies may not have resources available for the effort. Accordingly, unless Geospatial One-Stop devotes more resources to working with agencies on generating and posting metadata, its objective of creating an inventory of federal data holdings may be delayed. Finally, despite the creation of the Board of Directors, questions have been raised about the breadth of participation in Geospatial One-Stop. The chair of the board acknowledged that the small group of nonfederal representatives on the board may not be able to speak for all the states and thousands of local governments.

REFERENCES

Berry, J.K. (1993) *Beyond Mapping: Concepts, Algorithms and Issues in GIS*. Fort Collins, CO: GIS World Books.

Bolstad, P. (2005) *GIS Fundamentals: A first text on Geographic Information Systems, Second Edition*. White Bear Lake, MN: Eider Press, 543 pp.

5

GIS in Agricultural Production

GIS offers farmers various data management opportunities to analyze ways to increase production, reduce input costs, and manage farmland efficiently. From handheld computer mapping in the field to the scientific analysis of production data at the farm manager's office, geography plays a part by tying together disparate data that was previously never brought into one computing environment for analysis.

Data Collection in the Field

Farmers have great insight into how to best manage a farm's assets. While never being able to replace such expert knowledge, technology is certainly able to assist in the planning and implementation of important decisions related to farm management. In order to make the best decision, farmers need to have the most information at hand, often in the field. Now, new technology in computing allows the farmer to take the power of GIS into the field. ArcPad™ software can be used on handheld computers in the field for the creation, visualisation, and querying of data. Analyzing the data lets the farmer map relevant information such as crop stress from pest infestations, nutrient deficiencies, and water shortages.

Desktop Data Analysis and Interpretation using ArcGIS

Comparing different layers of agronomic information simultaneously is often enough to identify relationships that exist between seemingly unrelated data types. An example would be the location of weeds within a field and historical crop rotation data. Applying sound or advanced geostatistical tools to the farm operator's geospatial data results in optimum yields and lowest crop input investment. Identifying this spatial relationship supports better decision making for improved management or field treatments using variable rate technology (i.e., only applying field inputs where necessary), saving costs and reducing environmental impact.

GIS FOR AGRIBUSINESS

ESRI offers a scalable enterprise GIS solution that delivers improved information distribution between departments within an organisation using multiple software products. ArcGIS, ArcSDEArcIMSand ArcLogistics™ Route are fully compatible and can be connected via a company network.

GIS-Based Asset Management for Agricultural Businesses

Crop input manufacturers use spatial demographics and market demand models based on existing sales information and factors of influence to predict future manufacturing and marketing requirements. An underlying principle of successful business is acquiring knowledge related to the whereabouts of assets and demands, analyzing this information, and using it to provide an optimised strategy to deliver where existing and future business opportunities are located. Businesses that have implemented GIS are benefiting from increased supply chain efficiencies through linear network functionality. Reduced product storage times, better delivery routing and scheduling, and up-to-date asset management all work to release asset investments and return profits faster.

Using GIS in the Food, Feed, and Fiber Processing Industry

Out of necessity, processing companies manage their operations in fine detail by contracting farmers for delivery of specific quantities on specific dates. This management aims to reduce surplus processing assets or market demand that outstrips production capacity. Business models that are built containing natural phenomenon, without consideration of variations outside of the control of normal production practices, are significantly at risk. The food industry is no exception and frequently suffers as a result of disease, climate, and drought. Understanding of these potential risks is based on statistical analysis of historical information in a spatial or location-specific context. The Geostatistical Analyst extension provides agribusinesses with the tools to calculate these risks and better forecast any eventualities when they do occur.

IMPLEMENTING GIS IN GOVERNMENT

Federal, state, and local agricultural agencies have a broad responsibility to a wide range of seemingly disparate groups. All the way through the food supply chain from farmer to consumer, agricultural ministries and institutions collect, collate, manage, and disseminate information. These activities will increase with growing public awareness of farming practices and the resultant legislation. If there is a common theme in new efforts to better manage agriculture, it is communication, not only internally but also between government and consumer.

Regulatory Compliance Programmes for Crop Nutrients, Pesticides, and Manure

Crop inputs take many forms, but pesticides, manure, and phosphates are most hazardous to the environment. For this reason, state agricultural agencies must carefully manage and control pesticide applications to achieve an acceptable balance between the economic interests of the farming community and health concerns. Regulations on pesticide use vary from state to state, but all states require information to be collected and disseminated. Such information includes which, how much, and where pesticides have been applied. Since much of this information is tied to a discrete land parcel unit or field, it can be stored and accessed spatially. By storing this information, a GIS allows state officials to electronically manage initial permit application, monitor water quality, and control point source pollution events.

Disease Tracking, Pest Detection, Monitoring, and Control

Integrated pest management relies on accurate data related to the quantity and location of pests within a county. Using insect traps and field observations, government entomologists highlight areas where agricultural pests are apparent and recommend remedial measures. Whether on handheld units running ArcPad or the office desktop with ArcViewGIS plays an important role in data collection, interpolation, and analysis. Using ArcIMS, government officials can send up-to-date information on pest locations, areas of potential risk, and recommendations for eradication. A public Web site allows users to obtain real-time information, including maps of current pest outbreaks, that assists them in reducing disease contagion and its impacts.

Agricultural Subsidy Administrative Procedures Facilitated by GIS

Producers in many countries rely heavily upon subsidy payments to assist in maintaining viable commercial operations. Annual operational costs for such programmes are huge, and methods to improve efficiency by even 1 percent can have improved economic benefits. Subsidy claims have four distinct phases. The farmer first needs to be provided with the necessary information if the application is to be processed smoothly. Forms are then completed and submitted to the claims office. The claim is processed and verified, sometimes using satellite imagery but more often using field inspectors. Once the verification is complete, subsidy payments are released. Web-enabled GIS facilitates this work order flow process. Government agencies can now reach an increasing group of agricultural subsidy claimants via the Web. Subsidy applications can be completed online using ArcGIS technology.

ARCGIS TOOLS FOR AGRICULTURE

ArcGIS is a family of software products that forms a complete geographic information system for the fully integrated land information agency or a stand-alone system for the individual farm. ArcGIS is built on open, recognised standards and fully integrates with

IBM® DB2InformixSQL Server and Oracle® RDBMS. ArcView, ArcEditorand ArcInfo™ comprise the ArcGIS Desktop, while ArcSDE and ArcIMS make up the application services products.

ArcView

ArcView is an exceptional stand-alone desktop GIS product, as well as the entry point into ArcGIS, and provides core mapping and GIS functionality. ArcView provides geographic data visualisation, query, and analysis of most factors affecting agriculture land and improvement value. Land use, soil, and flood maps are just some of the geographic data that can be combined with parcel maps in ArcView to establish, adjust, and defend values for an area.

ArcEditor

ArcEditor includes all the functionality of ArcView and adds the power to edit topologically integrated features in a geodatabase or coverage. Other functionality includes support for multiuser editing, versioning, custom feature classes, feature-linked annotation, and advanced data types and geometry such as survey measures and coordinates. ArcEditor, with its advanced, easy-to-use drafting tools, is the ideal solution for multiuser enterprises.

ArcInfo

ArcInfo is the complete GIS data creation, update, query, production, and analysis system. It includes all the functionality of ArcView and ArcEditor and adds advanced geoprocessing and polygon processing as well as the classic ArcInfo applications and capabilities, such as Arc, ArcPlotand ArcEditthat make ArcInfo the de facto standard for GIS. Land records agencies find ArcInfo to be the complete GIS for converting and adjusting data from other departments, as well as state and national agencies, into their information.

ARCGIS EXTENSIONS

The architecture of ArcGIS supports a series of extensions, which dramatically extends the functional capabilities of the ArcGIS Desktop products for agriculturists. One of the key features of the ArcGIS extensions is that they operate seamlessly with ArcGIS Desktop software. This significantly reduces training, operating, and acquisition costs.

ARCGIS SERVERS

ARCSDE

ArcSDE is the gateway for GIS data in an agriculture geospatial DBMS. ArcSDE enables an organisation to move from the traditional approach of managing collections of geographic data files to an integrated environment in which all spatial data is modeled and managed as a continuous database in a DBMS. ArcSDE provides support for storing and managing all raster data formats supported by the ArcGIS Desktop applications. ArcSDE serves data directly to the ArcGIS Desktop products and through ArcIMS to a variety of lightweight Internet clients for both Internet and Intranet applications.

ArcIMS

ArcIMS is the foundation for distributing and disseminating GIS data on the Internet. ArcIMS enables a common foundation for the exchange and sharing of resources by providing a collaborative environment for food and agricultural organisations to share information. ArcIMS works with a variety of clients including ArcGIS Desktop products, ArcExplorerArcPad, and Web browsers. ArcIMS has proven very popular for providing public access to GIS data maintained by various departments. Ready access to this data by the public as well as intensive users of geospatial data, such as the real estate industry, has resulted in significant savings for public service departments.

MOBILE GIS

ArcPad

ArcPad software is a mobile GIS technology for field data collection and verification. ArcPad provides database access, mapping, GIS, and GPS integration to users in the field via handheld and mobile devices. The ability to take a portion of the agricultural geodatabase into the field, verify or edit the data, and then return the data to the geodatabase without additional processing is measurably increasing the efficiency of field consultants, farmers, and researchers.

HOW TO IMPLEMENT YOUR GIS

Data Requirements

Well-managed and documented databases are essential for successful GIS implementations. Data in mapped, image, and tabular form can be obtained from many sources. Agricultural departments may have the data already, find the data they need on the Internet, purchase data, or create their own data. Specific projects require specialised data that can be converted into a geographic format. Much data can be geocoded (i.e., given a locational value through an address, ZIP Code, or the like).

Hardware

Advances in computer hardware technology mean new options for agricultural GIS. Which option you select depends on many factors including the scale of the GIS implementation, the budget allowed, and the technical ability of the staff that will administer it. Options range from mobile devices, desktop computers, and workstations to centralised servers.

ESRI RESOURCES AVAILABLE TO THE AGRICULTURE INDUSTRY

GIS Solutions for the Farm and Agribusiness

This CD shows you solutions using GIS for those who work in the agricultural industry. Whether you are a farmer, research analyst, or crop manager or own an agribusiness, you will learn how GIS optimises your operations.

ESRI GIS Solutions for the Government in Food and Agriculture

The new ESRI GIS Solutions for Agricultural Government CD–ROM provides ideas and examples of where GIS is currently or will be used in state and local agricultural government activities. In addition, the CD contains free evaluation software, digital data sets of the United States, video presentations of ESRI software, recommended software configurations, and explanations about how a GIS can be implemented within a state or local government office.

TRAINING AND CONSULTING

Planning is essential to creating a successful GIS facility, starting with an understanding of the GIS business objectives and requirements. ESRI supports implementing an agricultural GIS through its Virtual Campus course named *Planning for a GIS*. The first module of this course is free and may be found at http://campus.esri.com. This Virtual Campus course was created by Roger Tomlinson, Ph.D., one of the world's most recognised authorities in GIS. This course lays out a proven planning methodology applicable for department wide or enterprise wide GIS projects. ESRI Professional Services has more than 30 years of application development experience in creating a plan of implementation for government departments.

REFERENCES

Chang, K. (2007) *Introduction to Geographic Information System, 4th Edition*. McGraw Hill.

Coulman, Ross (2001 - present) Numerous GIS White Papers

de Smith M J, Goodchild M F, Longley P A (2007) Geospatial analysis: A comprehensive guide to principles, techniques and software tools".

Elangovan,K (2006)"GIS: Fundamentals, Applications and Implementations", New India Publishing Agency, New Delhi"208 pp.

Harvey, Francis(2008) *A Primer of GIS, Fundamental geographic and cartographic concepts*. The Guilford Press, 31 pp.

6

Functioning of Geographic Information Systems

There are many ways to describe a Geographic Information System. Here are three working definitions:

1. *The minimal definition:* A GIS is a hardware/software system for the storage, management, and (with hardcopy or screen graphic) selective retrieval capabilities of georeferenced data. Definitions like this one are often used by vendors and users of vector-only GIS, whose objective is sophisticated management and output of cartographic data.
2. *A parallel definition:* A GIS is a hardware/software system for managing and displaying spatial data. It is similar to a traditional Data Base Management System, where we now think in *spatial* rather than in tabular terms, and where the "report writer" now allows output of maps as well as of tables and numbers. Thus we can consider a GIS a "spatial DBMS" as opposed to traditional "tabular DBMSs." Few people use this definition, but it might help to explain GIS to a DBMS user.
3. *A more aggressive definition:* A GIS is a system of hardware, software, and data that facilitates the development, enhancement, modeling, and display of multivariate (e.g. multilayered) spatially referenced data. It performs some analytical functions itself, and by its analysis, selective retrieval and display capabilities, helps the user to further analyze and interpret the data. Properly configured, the GIS can model (e.g. synthetically recreate) a feature or phenomenon as a function of other features or phenomena which may be related—where all features or phenomena are represented (characterised) by spatial and related tabular data. The analytical objectives described here are sometimes controversial—and often given lip service by cartographic GIS specialists who have not yet seen what can be accomplished scientifically by a select few GISs that go beyond cartographic approaches.
4. Another definition can be found at *the University of Edinburgh.*

WHAT IS GRASS?

Geographic Resources Analysis Support System (GRASS) is a public domain raster based GIS, vector GIS, image processing system, and graphics production system. Created by the US Army Corps of Engineers, Constriction Engineering Research Laboratory (USA/CERL) and enhanced by many others, it is used extensively at government offices, universities and commercial organisations throughout the world. It is written mostly in C for various UNIX based machines. Linux is one of its more robust implementations. GRASS contains over 40 programmes to render images on monitor and paper; over 60 raster manipulation programmes; over 30 vector manipulation programmes; nearly 30 multi-spectral image processing manipulation programmes; 16 data management programmes; and 6 point file management programmes.

GRASS' strengths lie in several fields. The simple user interface makes it an ideal platform for those learning about GIS for the first time. Users wishing to write their own code can do so by examining existing source code, interfacing with the documented GIS libraries, and by using the GRASS Programmers' Manual. This allows more sophisticated functionality to be fully integrated within GRASS. Other strengths include GRASS' pioneering of mixed resolutions in a data base, mixed geographic coverage areas in a data base, raster image compression techniques via run-length encoding and reclassification lookup tables, GRASS' rescaling of display images on the fly to fill the display screen, plus its fundamental design criterion of powerful computer-assisted scientific analysis of environmental issues (as opposed to merely going for intricate cartographic output of relatively simple processes). GRASS is usually supplied as free, non-copyright source code to be compiled on host machines. Some compiled binaries are also easily obtainable at no cost via the Internet. It runs on a variety of UNIX platforms.

In the early 1980s the U. S. Army Corps of Engineers' Construction Engineering Research Laboratory (USA/CERL) in Champaign, Illinois, began to explore the possibilities of using Geographic Information Systems to conduct environmental research, assessments, monitoring and management of lands under the stewardship of the U. S. Department of Defense. Part of the motivation for this action was new responsibility for the environment encoded into the National Environmental Policy Act of the late 1970s. Bill Goran of USA/CERL conducted a survey of available GISs, assuming that he could find several systems capable of environmental analysis, from which he could select one or more to recommend for use by CERL and perhaps others in the Department of Defense. However, he was surprised to find no GIS that satisfied his needs. What started as a selection process turned into a design exercise for his own GIS development programme.

USA/CERL hired several programmers, and began by writing a hybrid raster-vector GIS for the VAX UNIX environment. This made the team one of the first to seriously develop GIS for UNIX. Though they still faced challenges with different versions of UNIX, they developed procedures of coding in ANSI standard UNIX, avoiding "tweaking" the code toward any particular vendor-specific flavor of UNIX. GRASS developed a programming style characterised by:

1. Use of UNIX libraries where possible, plus the creation of GRASS libraries for repeated GIS-specific acts such as opening raster files that might be compressed (by run-length encoding) or not.
2. The ability to handle both major GIS data types: raster and vector.
3. The favoring of raster data processing, as scientific analysis was easier to encode with raster (than for vector) data models.
4. The ability to handle raster grids of mixed grid sizes in the same data base. This was a departure from raster's image processing tradition of requiring identical (and perfectly registered) grid cell arrays in each and every data layer.
5. The ability to handle raster grids with different areas of coverage. Again, this was a departure from raster tradition of having all grids be identical in geographic coverage.
6. The ability to run-length encode raster data files, in order to greatly reduce file sizes of most files.
7. The separate structure of reclassification files. Such files merely contained a look-up table noting the previous and new classes. This is MUCH more compact than replicating the original grid with different numerical values. A reclassified file of a 100x100 km square area of 10 metre grid cells would be a few hundred bytes, rather than 100 megabytes of uncompressed 8-bit raster data.
8. The acceptance of de-facto standard data models. While competitors created cumbersome (and in many cases secretive) data formats, GRASS accepted the de-facto standard Digital Line Graph vector format and unheaded binary raster grid format. GRASS later abandoned DLG as its internal vector file format, and let its raster format evolve. However, DLG and the unheaded binary raster grid are still routinely handled formats for GRASS, and its new formats are as open as its previous ones.
9. GRASS code was managed in several directories. Initial contributions were placed in the src.contrib directory. More solid code was moved to the src.alpha directory. After remaining in the src.alpha for one full release cycle, the code, with resultant bug fixes, moved to the most honorable level, the src directory.

GRASS was overseen by three levels of oversight committees. USA/CERL kept the ultimate responsibility for GRASS. It implemented most GRASS development, and carried out the day-to-day management of GRASS testing and release. The GRASS

Interagency Steering Committee (GIASC), comprised of other Federal agencies, met semi-annually to review development progress, and evaluate future directions for GRASS. (Academic and commercial participants in GRASS also attended GIASC meetings; only part of each meeting was "Federal-Agencies-only." GRASS eventually became nominally and officially a "product" of the GIASC, though everyone recognised USA/CERL's leadership role. The GRASS Military Steering Committee met periodically to review the progress of GRASS in serving its original intent: meeting the Department of Defense's needs to evaluate and manage the environment of military lands. The public interacted with CERL and GIASC through USA/CERL's GRASS Information Center. GRASS Beta testing was very widespread, and quite intensive for the leading users of GRASS. Several leading users, such as the National Park Service and the Soil Conservation Service, selected GRASS as its prime or only GIS. They made significant commitments to enhance and test GRASS, yet considered this investment well worth their while. They said that they had more influence over the direction of GRASS than they would over any known alternative system. They also felt that, despite their major efforts and expenses in supporting GRASS, they had a bargain in relevant power for the dollar.

Several universities adopted GRASS as an important training and research environment. Many conducted short-courses for the public, in addition to using GRASS in their own curricula. Examples of such leading academic users of GRASS are Central Washington University, The University of Arkansas, Texas A & M University, The University of California at Berkeley, and Rutgers University. Though GRASS received some criticism (some say) for being so good and so public, it was also reputedly borrowed from liberally by some developers of other systems. Though the first group might have viewed it as unfair competition, the second group may have noted that it was not copyright, and was a valuable testbed for GIS concepts. GRASS received an award from the Urban and Regional Information Systems Association (URISA) for quality software in 1988. As CERL and GRASS evolved through the late 1980s and early 1990s, CERL attempted to cut overhead costs associated with supporting the public domain version. It created and initially funded the Open GRASS Foundation, in cooperation with several of the leading users of GRASS. The Open GRASS Foundation has since evolved into the Open GIS Consortium, which is aiming for more thorough interoperability at the data and user interface level, but appears not to be taking advantage of the major open GIS testbed (GRASS). In 1996 USA/CERL, just before beginning the beta testing for GRASS version 5.0, announced that it was formally withdrawing support to the public. USA/CERL announced agreements with several commercial GISs, and agreed to provide encouragement to commercialisation of GRASS. One result of this is *GRASSLANDS*, a commercial adaptation of much of GRASS. Another result is a migration of several former GRASS users to COTS (Commercial Off-The-Shelf) GISs. However, GRASS' anonymous ftp site contains many enhancements to the last full version 4.1 release of GRASS. Many organisations still use GRASS feeling that, despite the lack of a major release in five years, GRASS still leads the pack in many areas.

A Re-Invogorated GRASS Management Model

In late 1997, a group at Baylor University took the lead in developing a new Website for GRASS. This quickly developing Website contains GRASS 4.1 source code and Sun Solaris binaries, GRASS 4.1 documentation, and an on-line manual. By November 1997 this site posted the first version of GRASS 4.2 source code and binaries currently for Sun Solaris) with Linux and Windows NT under consideration). GRASS 4.2 incorporates several enhancements from the CERL website, plus some of Baylor's own enhancements. Documentation for GRASS 4.2 is appearing; the group encourages cooperation in further development of GRASS, and is looking for partners. It hopes to use increased use of the World Wide Web in developing and managing GRASS. GRASS 5 development and compilation is underway. The site also links to the Blackland GRASS site at Texas A & M University, for those desiring very inexpensive access to GRASS for Windows 95.

Continued Assessment of Future GRASS Management

Note: An ad-hoc group is exploring the basic issue of continued, reconfigured, yea perhaps increased, value of GRASS as a public test-bed for GIS technology. It is exploring shepherding the testing and release of GRASS5.0, and exploring possibilities for a more distributed management model for GRASS design and development. It is exploring the universe of public domain spatial data processing software (including geographic information and image processing systems), and perhaps tabular data base management systems. How can such knowledge be (1) optimised as an open, public test bed for such technology and (2) better used by the public? Might this involve a Linux management model, perhaps?

SYSTEM REQUIREMENTS FOR GRASS

Minimum requirements include:

1. 8 Mbytes of memory (of course, more is better..)
2. 100 Mbytes of free disk space
3. ~40 mb for executables,
4. ~40 mb for source code (which you can ignore if you merely install the Linux binaries)
5. ~? for data (the veritable bottomless pit can be filled with data, if you so choose)

GRASS runs on Linux kernel versions as old as 1.2.13. GRASS will run in text mode. However, for displays of data, you will need X. If you are still running a version of X, it will probably work with GRASS.

HOW TO ACQUIRE GRASS

GRASS used to be available on tape from various companies that signed distribution agreements with USA/CERL. These companies usually supported specific platform environments, such as Masscomp, Sun, DEC, Hewlett Packard, IBM (risc), PC (running some flavor of UNIX), and Macintosh (running AUX). Until recently, the flavors of UNIX working on PCs generally were too low-end, or required too much added programming support (e.g. programming drivers for high-end graphics boards like the Number Nine boards of several years back) to be stable or complete. However, with robust systems like Linux, this problem is history. Similarly, few people acquire GRASS on tape, though a few do on CD-ROM. The main way to acquire GRASS is to get it via anonymous ftp from:

1. The new site at *Baylor University* : As of the date of this version of the mini-HOWTO, Baylor has source code for GRASS 4.1 and 4.2, as well as Sun Solaris compiled binaries. Blackland GRASS for Windows 95/NT is linked to from this site. Baylor is considering its own Linux and Windows NT binaries, as well. You should be able to compile the Baylor source code under Linux yourself, using information in this mini-HOWTO.
2. *The traditional site* at *USA/CERL* or from mirrors cited at USA/CERL's website:

   The ftp location is: moon.cecer.army.mil

THE FUTURE OF GRASS?

1. USA/CERL's announced intention is to use GRASS and COTS (commercial off-the-shelf software) for internal uses, to leave the GRASS public web- and ftp-site on its system indefinitely, and to sign cooperative research and development agreements with three companies: (1) the Environmental Sciences Research Institute (ESRI), (2) Intergraph, and (3) Logiciels et Applications Scientifiques (L.A.S.) Inc. The first two agreements encouraged the incorporation of GRASS concepts into ESRI's and Intergraph's commercial GISs. The third encouraged the adaptation of GRASS' concepts and code into a new commercial GIS by L.A.S. L.A.S. also offered to encourage the continuation of a public domain GRASS, as a viable stand-alone system and as a potential source of new ideas and code for L.A.S.'s GRASSLAND. One observer noted that the first two agreements might be akin to someone signing Linux over to Microsoft. The same observer considers the experiment by/with L.A.S. to be an interesting possibility—an attempt to keep viable public domain and commercial versions of GRASS.
2. Some people believe that GRASS will wither without USA/CERL's central management. Some believe that the Open GIS Consortium will successfully guide industry into an open architecture that will benefit all developers and users. Others believe that OGIS' effort will lead to a cacophony of almost similar (but not quite interoperable) vendor-specific "standards," so the loss of GRASS as an open

development platform will be felt sorely.

3. Some people believe that developments on some campuses and other sites may result in those institutes keeping GRASS for awhile, but in non-standard forms. In short, GRASS will undergo "cell division" and lead to a cacophony of internally valuable, but externally unused, GISs.
4. Others hope that GRASS' previous management model under USA/CERL has left it ready for a new model. Perhaps:
   (a) Under a new mentor, such as NASA (which needs an open, powerful and scientific, GIS integrated with image processing system for its Earth Observing System).
   (b) Under a distributed management model... perhaps somewhat like Linux?
   (c) Perhaps a bit of a hybrid? Perhaps a Web-based effort could spawn a series of usenet discussion groups beginning with

   comp.infosystems.gis.grass, and evolving to:

   comp.infosystems.gis.grass.academics
   comp.infosystems.gis.grass.publicservice
   comp.infosystems.gis.grass.commercialvalueadded
   comp.infosystems.gis.grass.commercialdistributors
   comp.infosystems.gis.grass.programming
   comp.infosystems.gis.grass.users
   comp.infosystems.gis.grass.centralcommittee

Clearly the topics are a bit tongue-in-cheek. However, under this model, a Central Committee (including representation of academic, public service [government and nongovernmental organisations], commercial distributors and value added firms, programmers, and users) would guide overall grass development and testing. The other special interest groups would serve their user communities. Academics, for example, would involve GIS and GRASS education, but would also try to pull GRASS development in its direction. Value added commercial developers would serve their own interests, including trying to pull GRASS development in a direction that would help their businesses. Users would help each other learn GRASS, develop workarounds to bugs, etc. GRASS offers considerable potential for:

1. Use as a scientific, as well as a traditional graphically oriented GIS. Many GISs can make pretty maps. Many of those GISs cannot easily perform certain scientific analytical functions as easily or powerfully as GRASS. GRASS was designed and

developed in response to a perceived need for scientific GIS, specifically for environmental analysis, and the environmental management/protection of public lands. Incidentally, there is at least one Web-based GRASS version. *GRASSLINKS*, developed at *The University of California at Berkeley*, uses Web forms to submit commands to the server, which creates .gif-based display output, places the images into pages, and serves them up to the requester. More on that later.

2. Education. GRASS is easier to teach and learn than some other GISs. It is easier to modify (for those that want to learn GIS as computer science, rather than as "geography") than most other GISs that come without source code and treat the programme as a magical black box. And, of course, it is more affordable for the student of GIS than many other GISs.

3. Applications research and development. Many universities have used GRASS. Its available source code, easy modification, easy scriptability, etc., give it distinct advantages over some more closed systems.

4. Public Service. GRASS has been used as a scientific GIS for many public service applications. There is considerable value in continuing a robust GIS that can ba packaged with any UNIX workstation. There is considerably more value if that UNIX workstation universe can include Linux (but is not constrained only to Linux).

5. GIS research and development. For example—do you want to experiment with a different data model? Add it to GRASS!

6. Commercialisation. This document gives contact information for a commercial version of GRASS. That company (and perhaps others?) may welcome your help in enhancing/supporting their product.

REFERENCES

Sajeevan G (2006) *Customise and empower*, www.geospatialtoday.com: April 2006. 40 pp.

Thurston, J., Poiker, T.K. and J. Patrick Moore. (2003) *Integrated Geospatial Technologies: A Guide to GPS, GIS, and Data Logging*. Hoboken, New Jersey: Wiley.

Tomlinson, R.F., (2005) *Thinking About GIS: Geographic Information System Planning for Managers*. ESRI Press. 328 pp.

Wise, S. (2002) *GIS Basics*. London: Taylor & Francis.

Worboys, Michael, and Matt Duckham. (2004) *GIS: a computing perspective*. Boca Raton: CRC Press.

Wheatley, David and Gillings, Mark (2002) *Spatial Technology and Archaeology. The Archaeological Application of GIS*. London, New York, Taylor & Francis.

7

New Interfaces for Agricultural GIS

GIS has become an increasingly popular tool for resource inventory, management and evaluation in Agriculture. It also has played an important role in developing expert systems in various fields of Agriculture as an essential technology for an integrated decision-making assistant system (IDMAS). GIS applications generally consist of various types of map data and attribute data. Map data are mostly in a vector format, while attribute data are usually kept in a database format, such as Oracle™, Informix™, SyBase™, SQL server, etc. Vector data are difficult to keep in the same database system with attribute data because of their uniqueness.

DRAWBACKS IN GIS DISTRIBUTION

In spite of its popularity, it raises a couple of drawbacks in its distribution to end users as a part of expert systems. Along with the relatively high cost to purchase many copies of GIS systems for end-users, the release of system-specific database to end-users may give us another question in security, stability, consistency, and accuracy of database themselves. Furthermore it will demand enormous efforts from system developers and managers when the system needs upgrades on either GIS or database.

NEW INTERFACE FOR GIS

The obstacles to the extended use of GIS technology in Agriculture can be, at least partially, removed by employing client-server model with GUI capability. The newly emerging way of information exchange via network will be Web technology on the Internet/Intranet. The Web-oriented user interface can contribute to the remote operation of GIS applications, with which users can communicate interactively using GUI through distributed computing systems.

An intrinsic problem in Web interface development for GIS lies in the low transmit rate of Internet at present. The success of information exchange via Internet will rely on how these technologies can be integrated into the most efficient way to enhance information transmission between server and client. Depending upon the type and size of information considered, diverse Web technologies can be employed, though they are still under development or enhancement.

CURRENT TRENDS IN WEB

The Web functionality has progressed from simple document fetching to more interactive applications as follows: (a) fetching HTML or other formatted documents from fixed locations, (b) fetching documents from backend systems, such as databases, using the CGI, (c) building interactive systems using HTML forms and CGI, (d) using Java script to increase GUI capabilities, (e) using Java applets to provide client-side functionality, and (f) currently GUI capabilities combined with remote invocations.

Problems in CGI

The inter-activity of HTML-based interface through the CGI enables the execution of programmes on the server side, but unfortunately the CGI has a number of limitations and drawbacks. Writing a client as a sequence of HTML pages on URLs is an extremely tedious task. Data transferred from client to server must be parsed each time a new CGI call is received. There are a number of performance bottlenecks in the CGI-based approach. As a result of an invocation a complete HTML page is returned to the client side. These HTML documents contain a lot of repeated text and formatting data that remains unchanged since the last client action. The amount of unchanged HTML often outweighs the amount of actual data produced by the application programme by an order of magnitude. HTTP, the most popular protocol of the Web protocol suite, is not very efficient. The major performance bottleneck occurs because multiple connections can be created loading a single URL, and the connection management creates a significant performance overhead. Furthermore, the CGI will start a new operating system process each time the application processes a user input, and any server-side state must be read from persistent storage or communicated from another process.

Database Problems with Java

Java gives the developer some huge benefits for developing applications, but it also has some serious drawbacks. One of the drawbacks of Java is that it does not make developing GUI applications easy. Java limits the user to predefined managers for laying components on the screen. It does enable developers to place components on exact positions within the window, but which defeats the purpose of programming for multiple platforms. One of the main disadvantages of the Java™ 1.02 release was that it did not contain any support for accessing databases. This drawback limited the usefulness of Java in the business arena. However, the new release of Java™1.1 provides full support for databases through JDBC™.

To solve this problem of database access in Java, Sun has introduced Java Database Connectivity (JDBC). JDBC enables Java applications to connect to any database using various drivers and a new set of Java API objects and methods. The JDBC provides various objects for connecting to a database, executing SQL statements, and executing stored procedures. JDBC also provides various objects and methods for getting information about objects within the database.

Using JDBC Instead of CGI

Until now, the only way to access a database through Java has been to use streams in Java to call and access CGI programmes. Calling a CGI script from Java lets you call a separate programme that accesses the database and returns results. Using this approach is slow and also lets more bugs creep into your application. Developing in two different development languages requires the knowledge of two technologies. When you use JDBC you need to know only the Java language, whereas when you use CGI you must use Java and another language.

Another reason to use JDBC is that it is noticeably faster than the CGI approach. Using CGI usually requires that you call another programme that must be executed by the computer. This separate programme then accesses the database and processes the data, returning it back to the calling programme in a stream. This requires multiple levels of processing, which in turn increases wait time as well as enables more bugs to appear. Whereas calling a CGI script involves actually executing a new script, usually through a Web server, executing a JDBC statement against the database requires only some sort of server that passes the SQL commands through to the database. This speeds up dramatically the time it takes to execute SQL statements. Whereas the CGI script must actually connect to the database and process the results, the JDBC solution lets your Java application have the connection to the database so that the application can perform all of the processing.

Java/CORBA

There are two of the major object models used in distributed computing: Common Object Request Broker Architecture (CORBA) and Java. They each introduce a different approach to distributed computing. CORBA provides an infrastructure that enables invocations of operations on objects located anywhere on a network as if they were local to the application using them. Java introduces platform-independent, low-level code, which, when integrated with Web protocols and browsers, results in what are known as applets. In this approach, instead of invoking a method on a remote object, the code for the class providing the method is transferred across the network, run locally, and then the method is invoked on a local object instance.

The Java programming language does not directly support the development of distributed applications or systems. The only way to implement distributed applications that is directly supported in Java is to use the network library classes in the package java.net. Those classes provide an API for the handling of URLs and an API to UDP/IP and TCP/IP sockets. The Java language binding for OMG IDL provides an application programmer with CORBA's high level distributed object paradigm: (a) Interfaces defined independently of implementations, (b) Access to objects implemented in other programming languages, (c) Access to objects regardless of their location, (d) Automatic code generation to deal with remote invocations, and (e) Access to standard CORBA services and facilities.

WEB INTERFACES FOR GIS

Server-side Web Interface

To deal with information dissemination from GIS applications, CGI was employed to run ARC/View as a server side application for Web client on the Internet. It has an advantage in dealing with a relatively large size vector data, because only requested information has to be delivered to client after processed by a GIS application on the server side. Tested a simple CGI prototype that can execute Arc/View™ on Windows NT server™ with WebSite™ as a Web Server. After user made selections on themes on the Web browser, it was posted to Web server via CGI with Avenue™ that is an intrinsic script language of Arc/View. According to Avenue scripts in CGI message, Arc/View becomes executed following options chosen by end-users.

It will be a promising alternative where Internet is too slow to accommodate large size vector data at present. The image in WMF format produced from ARC/View according to users' request will be downloaded to the client's Web browser. The outcomes from Arc/View are transformed into WMF format, then transmitted to client side via Internet. Any browsers with an appropriate plug-in can display this WMF file. When the next version of Arc/View can generate GIF/JPEG file, it will become simpler in displaying the outcomes from GIS application because it will not be necessary transforming into WMF format. However, when asked from many users simultaneously, it will overload server's processor, then it will take much time to meet every demand from endusers. It also lacks of the direct inter activity for a map that is a common and important function in stand-alone GIS applications. This approach, thus, needs a supplement to facilitate the bi-directional communication between GIS developers and end-users in the future. Promising ways of supplementing these problems will be client-side interactive interfaces by using ActiveX or JAVA applet™.

An Alternative way of CGI-based Web interface will be Java Servlet™ as Server-side interactive interfaces. Up until now, the Java story has largely been a clientside story. JavaSoft™ has introduced server technology that is written in 100% Pure Java™ and that is capable of being extended dynamically by loading compiled Java code known as a servlet. While applets provide a way of dynamically extended the functionality of client-side browsers, servlets let you dynamically extend the functionality of network servers. Servlets are to servers what applets are to browsers. Servlet code can be downloaded into a running server to extend its behaviour in order to provide new, or temporary, services to network clients. Servlets provide a Java-based solution used to address the problems currently associated with doing server-side programming, including non-extensible scripting solutions, platformspecific APIs, and incomplete interfaces.

Servlets are objects that conform to a specific interface that can be plugged into a Java-based server. Servlets are to the server-side what applets are to the client-side - object bytecodes that can be dynamically loaded off the net. They differ from applets in that they are faceless objects (without graphics or a GUI component). They serve as platform independent, dynamically loadable, pluggable helper bytecode objects on the server side that can be used to dynamically extend server-side functionality. When you use servlets to do dynamic content you get the following advantages: they're faster and cleaner than CGI scripts, and they provide all the advantages of Java (run on a variety of servers without needing to be rewritten). There are many features of servlets that make them easy and attractive to use:

1. can be loaded and invoked from a local disk or remotely across the network.
2. can be linked together, or chained, so that one servlet can call another servlet, or several servlets in sequence.
3. can be called dynamically from within HTML pages, using server-side include tags.
4. are secure—even when downloading across the network, the servlet security model and servlet sandbox protect your system from unfriendly behaviour.

Client-side Web Interface

In the context of client-server software solutions—including the serving and browsing of Internet web pages—Java applets have enabled client browsers to extend their behaviour by downloading compiled code from network servers. Applets have changed the nature and meaning of distributed computing. Applets have opened new worlds by enabling the convenient distribution of compiled code from a central server. A JAVA-enabled Web interface for GIS applications was tested with which user can communicate interactively through Internet. This interactive Web interface for GIS application offers typical map display functions like zoom, pan and drag as well as functions of displaying attributes. It primarily runs on the shape file format generated by ARC/View for map display. Shape files were concatenated to reduce its size and project files were modified to accommodate only necessary parts such as legend-related contents.

GIS Lab. of NIAST(RDA, Korea) has created a suite of national soil vector maps with attribute data for most of the southern Korean peninsula. They supplied vector and attribute data for Yangpyong watershed for a demonstration purpose of this study. Other sample data from ESRI has been also used to test this interface. The following give us a brief description on files and tools having been used to develop this Web interface. A new user interface for GIS applications was developed with which user can communicate interactively through Internet. Not only end-users can gain access to outcomes from map display being operated in GIS applications from any web browser, but also they can supply region-specific information to application developers through Internet. This interactive Web interface for GIS application offers typical map display functions like zoom, pan and drag as well as functions of displaying attributes. It primarily run on the shape file format generated by Arc/View for map display, while attribute data are accessed via independent D/B server through JDBC:ODBC interface.

Currently it only offers only limited interactive functions that are mainly related to the map display. Users can define classification, map adjustment like zooming, panning, and dragging, attribute analysis through sorting, charting, simple calculation. Shape files were concatenated to reduce its size and project files were modified to accommodate only necessary parts like legend related contents. DBF files were transformed to MS Access™ mdb format for ODBC systems database Any existing thematic maps in a shape format produced from Arc/View can be displayed with basic options like zooming, panning, dragging, overlaying etc. User can define own legend number, colors, labels for numeric attributes, then display according to his/her selection. For a specific attribute field, user can choose any record to identify its order in magnitude by using Quick Sort Algorithm. Frequency distribution with designated intervals can be shown for a specific attribute field. All field data are displayed with 20 records at once. In this dataset, user can conduct simple calculation within/among fields/records. User also can identify the specific region on the map with designated field value.

REFERENCES

American Society for Photogrammetry and Remote Sensing and American Congress on Surveying and Mapping, 1987, GIS for Resource Management: A Compendium, pp288

ESRI, 1995, ArcView Version 2 Shapefile Technical Description, ArcView White Paper Series.

ESRI, 1995, Understanding GIS, The ARC/INFO Method, Self-study workbook, pc-version.

Stephen R. Davis, 1996, Learn Java Now, Microsoft Press, pp. 353

Sams net, 1996, JAVA unleashed, pp. 958

ESRI, 1997, ESRI Live Internet Demonstrations (http://maps.esri.com/ESRI/esri.htm)

GeoSystems, 1997, MapQuest (http : // www. mapquest.com/)

Storm Cloud Co. 1997, WebDBC v3.0 - JAVA-ODBC interface and Web database developer

OMG, 1997, Java Programming with CORBA, pp425

B.L. Lee, Y.C. Kim and M.E. Park, 1997, Interactive web interface for GIS applications in Agriculture, Korea Agricultural Information and Technology, Vol.6 No.1, pp136-139

8

Remote Sensing in Farms

Remote sensing is "...concerned with the collection of data by a sensing device not in contact with the object being sensed, and the evaluation of the collected data, which is then termed information and is presented in map form or as statistics.". Clearly, the concept of RS covers a huge field—a field within science and technology which encompasses a vast "applications domain", i.e. in the sense of inputs of applied science, applications in the processing field and in the sense of the ways in which RS outputs can be applied. Since Butler et al have already outlined the historical growth of RS (for the FAO), we will confine our resumé of its development to a few key advances. The human eye is a remote sensor and, although we can capture images which may be stored in the brain and later retrieved, we can only reproduce them in a subjective sense. The eye can only capture visible radiation which occupies a very small part of the complete range of radiation (which is known as the electromagnetic spectrum). To overcome these deficiencies various instruments or systems have been invented and developed. We will be concerned here with those instruments or systems which capture data from an aerial perspective, i.e. allowing maps to be easily created.

As a means of capturing images, the camera was first developed in France in the 1830s, but it was not until 1858 that the first aerial photograph was taken from a captive balloon near Paris. During the rest of the 19th century advances were made in cameras and additional camera platforms were experimented with. The first photograph from an aeroplane was taken in 1909 over Centocelli in Italy. During the first World War aerial photography was utilised on a large and systematic scale, both in Europe and the Near East, with specially designed cameras and film processing techniques being developed. It was during this period that photo interpretation became a recognised field of expertise. Civilian use of vertical aerial photography greatly improved during the 1920s and 1930s because of advances in both photographic methods and in the aeroplane as a platform. Aerial photography was used in the compilation of topographic maps, by geologists, for esters and planners, mainly in north America and Europe but occasionally to acquire information from more remote areas which might otherwise be unobtainable. During World War II further developments occurred, e.g.:

1. The water penetration capability of aerial film was recognised which meant that bathymetric data could be acquired.
2. Colour infrared film was developed for camouflage detection.
3. Advances in radar technology permitted the development of smaller transmitting and receiving equipment, appropriate for airborne use.
4. A large area of the Pacific war zone was photographically mapped.

During the 1940s and 1950s large-scale, complete country coverages were undertaken using black and white panchromatic aerial photography, i.e. for many of the colonial countries, especially in Africa. By the 1960s aerial photography had been operative long enough to allow for the study of spatial/temporal variations in the environment. The period from the late 1950s has been extremely active for RS, with developments in the whole applications field occurring at an exponential rate. With satellite launches occurring regularly, following SPUTNIK 1 in 1957, the interest in RS concentrated on the use of this new and unique platform. In 1959 the first earth images were transmitted from EXPLORER 6 and the first meteorological satellite, TIROS-1, was launched in 1960. The next major advance for RS occurred with the launch of the Earth Resources Technology Satellite (ERTS-1, later renamed Landsat 1) in 1972. This was the first satellite designed to provide long-term, uniform global coverage, having the ability to transmit data gathered on a variety of instruments, for eventual mapping at scales of 1:250,000 to 1:1 000,000. Since Landsat 1 there has been a succession of earth monitoring satellites launched, first by the U.S.A. and the U.S.S.R., but more recently by other countries. Their equipment has become progressively more sophisticated allowing a greater range of imaged data to be interpreted at a more detailed spatial resolution.

As both Howard and Butler et al make clear, we should not let the blossoming satellite technologies mask the fact that airborne RS is still thriving and vital. Aircraft have a number of distinct advantages over satellites, mainly in terms of their flexibility of altitude, scheduling and payload. They do not have the same cloud cover problems that satellites have and they can provide low cost, excellent images of smaller areas. Howard estimates that airborne techniques, using colour infrared technology at high altitude, can allow more than 20,000 km^2 to be photographed, and be thematically mapped at a scale of 1:25,000 to 1:100,000, each day. It is the rapid surge in the electronics industry which has permitted the post 1960s boom in RS, i.e. because as well as providing for actual developments in data capture, data transmission, image processing, etc., it has spawned the computer technology advances which have been vital to all aspects of space science and have allowed the huge data streams to be efficiently handled. This surge in RS has also been aided by the influx of new ideas from a variety of related disciplines, by the availability of funding for space related activities and by the access to an increasing range of software and hardware. RS has a number of positive advantages over other sensing systems:

1. It allows change to be monitored in a systematic and orderly way.
2. It is efficient and very cost-effective in per km2 terms.
3. It overcomes many data collection problems, e.g. in isolated areas and the fact that normal data collection may terminate at political boundaries.
4. It can provide for instantaneous updating of information.

Jackson and Mason report that modern RS has now successfully overcome the problems which it had in the 1960s and 1970s of being "technology pushed". As useful applications of RS imagery have been developed, especially in the fields of food production and environmental awareness, RS is now also being "user pulled". This recent trend has been greatly helped by the growing ability to successfully integrate remotely sensed data into GIS—in fact, we would suggest that if it was not for the functionality offered by GIS, then the future for RS might be rather uncertain.

ELECTROMAGNETIC RADIATION—THE BASIS OF REMOTE SENSING

"The aim of environmental remote sensing is to utilise sensors, which are mounted on aerial platforms, to identify and/or measure parameters of an object according to variations in the electromagnetic radiation (EMR) emitted by, or reflected from the object." Contained within this statement are a number of concepts.

Although we cannot see light (or sound) travelling, we know that it does so. This travel involves the transfer of energy through space or matter in the form of wave motions. The waves that make up EMR travel at a constant speed of 300 million meters per second. All objects reflect and radiate EMR. The amount of electromagnetic energy emitted is a function of the object's temperature - as temperatures increase, the intensity of the radiation emitted increases. There are a whole family of waves which collectively are called electromagnetic vibrations and which vary in their wavelength, and hence (since their speed is constant in space) their frequency. This family may be displayed as a spectrum of energies, hierarchically arranged by wave frequency or wave length. The spectrum covers a vast continuum of wavelengths as indicated, and it is usual to rather arbitrarily differentiate between the major wave bands based on certain properties such as their source, method of generation, means of detection, selected applications, etc. Only specific wavelength bands are of interest to RS, i.e. those forming the "windows of transmission", because at these wavelengths filtering out of EMR by the atmosphere is at a minimum. These bands can be shown as in Table 1.

Table 1 Wave Bands of the EMR Spectrum of Interest to Remote Sensing

Wave Band and Wave Length	Detectors	Some Characteristics
Visible 0.4–0.7 um	Black & white plus colour photography. T.V. camera. Optical scanner.	High atmospheric scattering effect. Most EMR is reflected solar radiation therefore only used in day-light. Penetrates water.
Near Infrared 0.7–3.0 um	Infrared scanner. Infrared thermography. Multi-spectral scanner.Photo-graphy. Optical scanner.	High reflectance of vegetation. Again solar energy reflected by surfaces.
Medium Infrared 3.0–8.0 um	as above.	as above.
Thermal 8.0–1000 um	Radar. Side Looking Air-borne Radar (SLAR). Optical scanner.	Predominantly radiation emitted by the earth and atmosphere. Does not penetrate clouds.
Microwave 1 mm–100 cm	Radar. SLAR. Scanning radiometer.	Can penetrate clouds. Imagery acquired in active or passive mode—daytime or night time.

The energy which is sensed by the different RS systems is a function of various parameters which might affect the energy before it is received by the sensors. This is shown in Figure 1 which indicates that EMR can be natural, either reflected light and other radiations from the sun (Source 1) or emitted heat from the earth (Source 2), or it can be man-made such as from a power station or a radar system. The amount and type of radiation emitted or reflected depends upon incident energy (mainly from incoming solar radiation), the nature of the earth's surface and on the interaction with the earth's atmosphere.

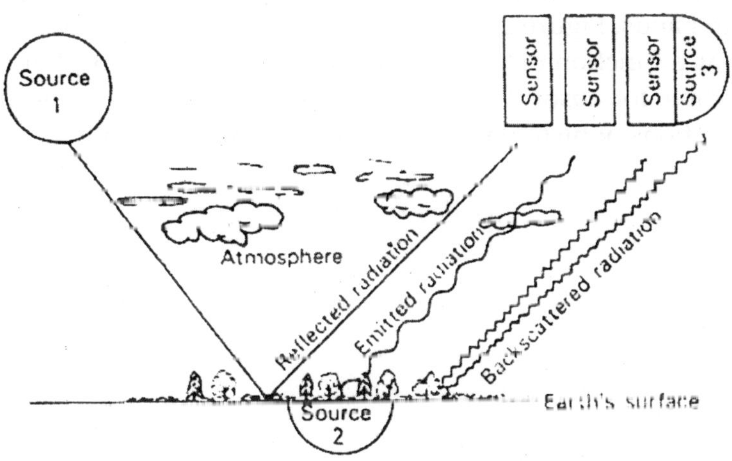

Figure 1: The Key Features of the Remote Sensing Data Collection Process.

Incident Energy

This comes mainly from the sun and, in the range of the visible and near infrared part of the spectrum, it is the proportion of the incident energy reflected by the "object" on the ground. When the energy sensed is in the range of thermal radiation it comes mainly from the emission of the "object" on the ground, which is itself a function of the sun's incident energy which has been absorbed by that object and then re-emitted as thermal radiation. Incident energy from the sun will vary with season or latitude (affecting the angle of the sun), with the length of time the sun has been shining and with the angle of the object on the ground. When analyzing remotely sensed data it is important to consider dates, time of acquisition and relief.

Effects of Atmosphere

The atmosphere can affect the amount of radiation received by the sensor because the atmosphere itself is heterogeneous, being made up of many gases as well as having dust particles and other pollutants. The atmosphere may scatter light in the visible band and absorb it in the ultraviolet and infrared bands. About 18% of the incident radiation in the atmosphere is absorbed or scattered and about 35% of the incoming solar energy is reflected by the earth and the atmosphere, including clouds. Scattering is caused by particles in the atmosphere reflecting the energy, and the intensity of the scattered EMR depends on the ratio of the wavelength to the size of the particles. Scattering caused by small particles is selective relative to the wavelength, affecting shorter wavelengths more; scattering related to large particles is non-selective, affecting all wavelengths. Because of scattering, the energy received by the sensor includes reflections from the atmosphere as well as from the target (object). Complex algorithms are needed to correct this effect. Atmospheric absorption reduces the amount of EMR reaching the sensor in some wavelength bands. Figure 2 shows the percentage of EMR which can pass through the atmosphere as a function of wavelength (revealing the atmospheric windows), and the gases responsible for absorption are noted. Microwave radiations are unaffected by atmospheric conditions, which makes them very useful, especially in cloudy areas such as the tropics.

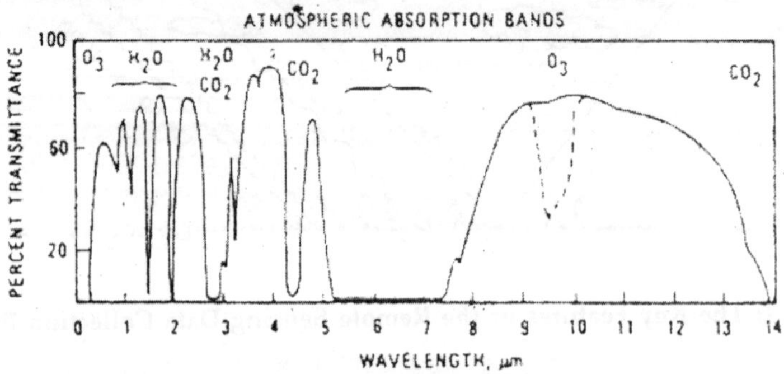

Figure 2: Percentage of EMR Able to Pass Through the Atmosphere as a Function of Wavelength.

Ground Influences

The element of the earth's surface which is in the field of view of the sensor (the target or object) will produce, by reflection or emission, the EMR measured by the RS sensor. The amount of energy transmitted or reflected depends on what the target consists of and the thickness of it. A target also absorbs radiation and this can affect the temperature of the target and thus the amount of energy radiated per second. So all targets (or objects) in the environment emit and reflect different intensities and types of EMR in different portions of the spectrum, i.e. they have a so-called spectral signature which is predictable and repeatable. Figure 3 gives the spectral signature of various natural features. These signature curves are dependent on a number of interactions between incoming radiation and the micro and macro-structure of the matter irradiated. Spectral signatures may vary temporally, e.g. as plants grow, or spatially with different types of vegetation, different soil conditions, water availability, effect of topography, etc.

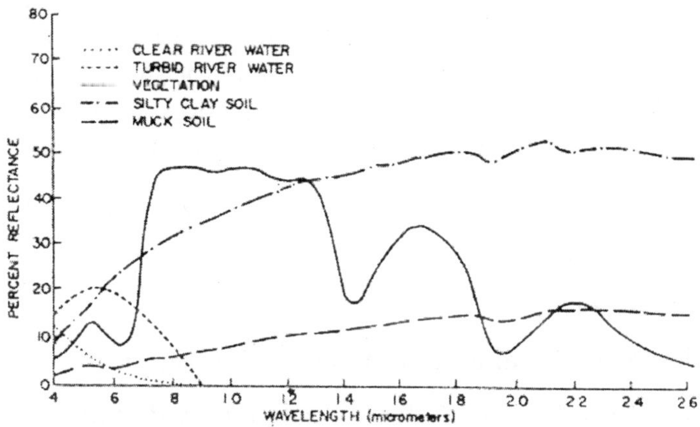

Figure 3 Spectral Signature of Various Natural Surface Features

REMOTE SENSORS

Sensors are the devices used to gather EMR. They will typically consist of four components, i.e. collectors, detectors, signal processors and recording units. There are several ways of classifying sensors given below:

1. *Framing systems*: These include various types of camera which record instantaneously an entire image.
2. *Scanning systems:* These employ a detector (electronic sensor) which sweeps across a scene in a series of parallel lines collecting data in order to record an image. They may employ passive sensors, which record reflected or emitted EMR from natural sources, or active sensors which illuminate an object with their own radiation source, and then record the "echo".

Framing Sensor Systems

Cameras may be used from various platforms. The still photography camera is the best known, simplest and cheapest of all sensors, and still photography produces images having a better resolution than those produced by electronic sensors. Cameras may produce simple, single images in one spectral band which are suitable for many referencing purposes, or they can produce overlapping pairs of aerial photographs which, when viewed using a stereoscope, give a three dimensional perspective of the landscape. But their chief use is in multi-spectral photography. Here a number of cameras may take simultaneous images of an object, using several band-pass filters, which each allow EMR information relative to particular wave bands to be recorded. The number, position and width of suitable colour filters can be optimised in a problem oriented manner so that the controller is able to discriminate between a wide range of features (within the visible and infrared bands—in the wavelengths between 0.4 and 1.3 um). The quality of the photographic image will depend on the inter-related factors of: focal length; the angle of view; scale of the photograph; the contrast ratio; the picture resolution and the film speed. There are a variety of cameras suitable for satellite and/or aerial RS, and their selection depends upon the nature of the application. In principle mapping frame aerial cameras are similar to normal cameras except that they have:

1. Calibrated lenses.
2. High geometric accuracy.
3. A medium to large format.
4. A more complex mechanical and electrical configuration.

Though cameras have the advantages already noted, they do have some disadvantages, i.e. there is a loss of resolution during photo-chemical processing or copying, or in analogue digital conversion for subsequent computer processing, and they have a limited spectral sensitivity. They can also only function in favourable weather conditions.

Framing System Films

Cameras will require different types of film for different purposes. The main types, according to their range of spectral sensitivity, are:

1. *Orthochromatic:* These have a very good discrimination in the green bands and are used mainly for cartographic reproduction.
2. *Panchromatic:* These cover the whole of the visible spectrum, with good sensitivity, except for green bands (0.5 um) which can be compensated for with a filter. This film is inexpensive, is easy to process, has a high spatial resolution and special filters can be used to enhance selected objects (targets).

3. *Black and white infrared:* This is similar to panchromatic except that its greater spectral sensitivity means that near infrared wavelengths can be recorded in addition to visible light. Usually a dark red filter is used to screen out the visible portion of the spectrum, so that only the near infrared portion is recoreded, which results in a greater penetration of the atmosphere. This film is used mainly for detecting different vegetation stages and types, plus the existence of water.

4. *Natural colour:* The spectral range of this film is similar to that of panchromatic. It is composed of three layers, each sensitive respectively to the three primary colours-blue, green and red. Colour images offer a range of about 20 000 natural shades whilst black and White is limited to only 200 grey tone shades, i.e. Colour film allows the distinguishing of many more features, because of its greater sensitivity to tints and shades. Colour film has a number of applications, e.g. its sensitivity to sub-surface water makes it especially useful for coastline definition and the estimation of water depth and sediment content. However, colour photographs are more expensive, they have a less good image definition and they cannot be taken from a high altitude.

5. *False colour:* These films are formed by having three layers sensitive to green, red and near infrared radiations, respectively modulated into blue, green and red. The film has been moderated to achieve several advantages, e.g. a high penetration of the atmosphere, sharp resolution and definition of water bodies and a good response to the infrared reflectivity of healthy vegetation. However, these films have a limited exposure tolerance and the film requires refrigerated storage.

Scanning Sensor Systems

This is the main alternative to photographic systems for detecting and recording EMR. Several bands of the EMR spectrum, either from the ultraviolet to infrared regions (multispectral scanners) or the microwave bands (radiometers), may be scanned simultaneously by optically splitting the collected radiation and diverting each part to a separate detector element. Final image products can be photographs or computer compatible tapes containing digital data. Scanning sensors can be either passive or active. Passive sensors detect natural incoming EMR and active systems detect system-generated EMR (so called "echo").

Passive Sensors

These sensors are called radiometers and they can detect EMR within the ultraviolet to microwave wavelengths. Two important spatial characteristics of passive sensors are:

1. Their "instantaneous field of view" (IFOV)— this is the angle over which the detector is sensitive to radiation. This will control the picture element (pixel) size which gives

the ground (spatial) resolution of the ultimate image (Figure 4), i.e. the spatial resolution is a function of the detector angle and the height of the sensor above the ground. For more details on spatial, spectral, radiometric and temporal resolutions see Lechi.

2. The "swath width"—this is the linear ground distance over which the scanner is tracking (at right angles to the line of flight). It is determined by the angular field of view (AFOV—or scanning angle) of the scanner. The greater the scanning angle, the greater the swath width (Figure 4).

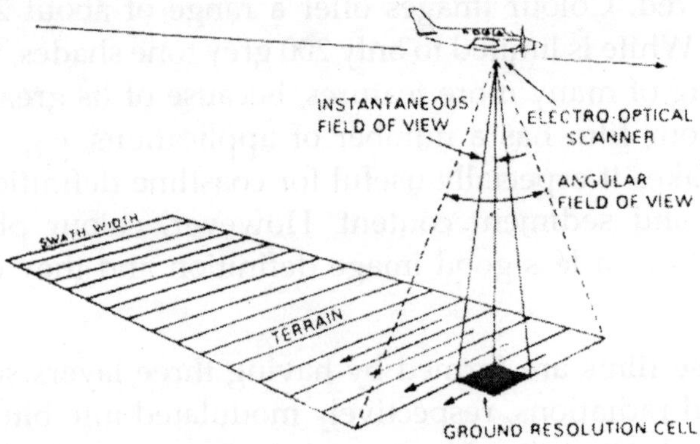

Figure 4: The Concept of IFOV and AFOV.

There are two main categories of passive sensor:

1. *A mechanical scanning radiometer:* This is an electro-optical imaging system on which an oscillating or rotating mirror directs the incoming radiation onto a detector as a series of scan-lines perpendicular to the line of flight (Figure 5). The collected energy on the detector is converted into an electrical signal. This signal is then recorded in a suitably coded digital format, together with additional data for radiometric and geometric calibration and correction, directly on magnetic tape on board the sensor platform.

2. *A push broom radiometer:* This uses a wide angle optical system in which all the scenes across the AFOV are imaged on a detector array at one time, i.e. there is no mechanical movement (Figure 6). As the sensor moves along the flight line, successive lines are imaged by the sensor and sampled by a multiflexer for transmission. The push broom system is generally better than the mechanical scanner since there is less noise in the signal, there are no moving parts and it has a high geometrical accuracy.

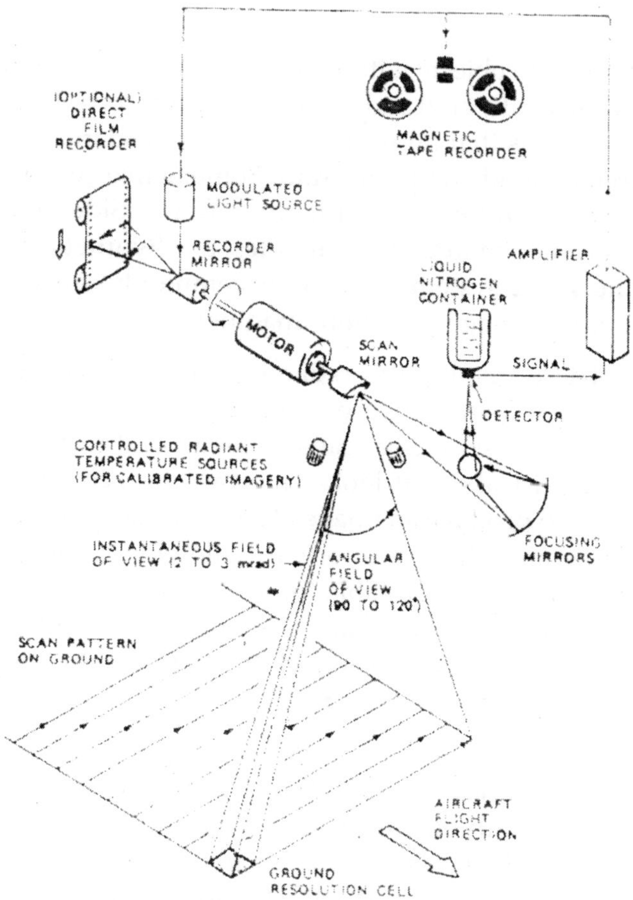

Figure 5: Optical Mechanical Scanning System

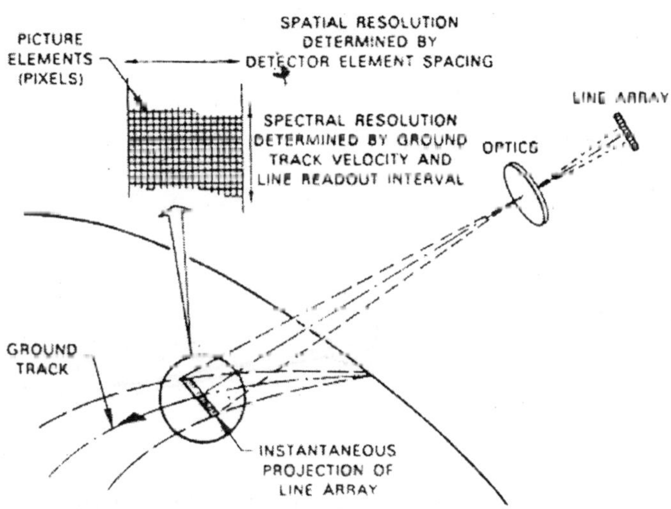

Figure 6: Characteristics of a Push Broom Radiometer

Active Sensors

All active sensors illuminate objects with their own source of radiation. The illumination will either induce an object to emit radiation or cause it to reflect the sensor produced radiation. This ability means that no sunlight is required so imagery can be recorded by day or night or through clouds and light rain. Some active sensor systems are surface-based, e.g. sonar, others could be carried in aircraft, e.g. Side Looking Airborne Radar (SLAR) whilst others can be mounted in satellites, e.g. Synthetic Aperture Radar (SAR). Airborne and satellite active systems, which are commonly called Radar, and which are generally classified either imaging or non-imaging:

1. *Imaging Radars:* These display the radar backscatter characteristics of the earth's surface in the form of a strip map or a picture of a selected area. A type used in aircraft is the SLAR whose sensor scans an area not directly below the aircraft, but at an angle to the vertical, i.e. it looks sideways to record the relative intensity of the reflections so as to produce an image of a narrow strip of terrain. Sequential strips are recorded as the aircraft moves forward allowing a complete image to be built up (Figure 7). The SLAR is unsuitable for satellites since, to achieve a useful spatial resolution, it would require a very large antenna. A variant used in satellites is the SAR whose short antenna gives the effect of being several hundred times longer by recording and processing modified data.

2. *Non-imaging Radar*: These are also called scatterometers since they measure the scattering properties of the region or object being observed, i.e. the roughness of the surface over a wide swath on either side of the spacecraft. A type of scatterometer is the radar altimeter which can provide an accurate height assessment for satellites, and these measurements can yield valuable topographic or sea surface roughness variations.

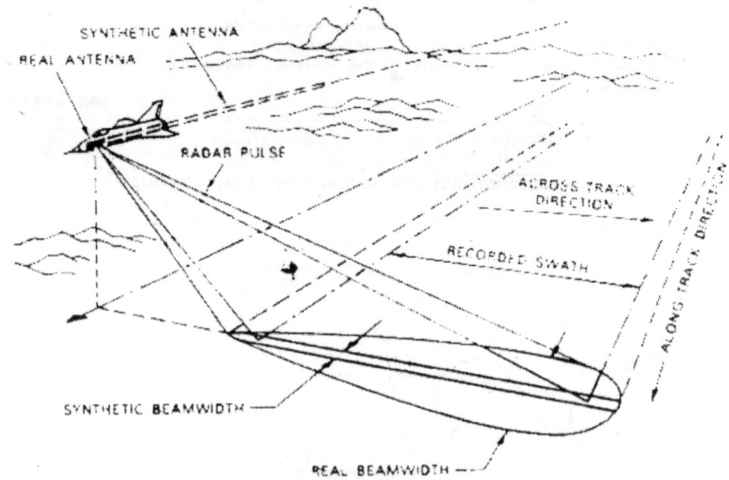

Figure 7: The Synthetic Aperture Radar System.

A further type of active sensor is the laser radar (LIDAR). LIDAR s use lasers to generate short, high power light pulses. These can be used to measure the intensity of light back-scattered by the target as a function of the distance from the sensor. Because of their size LIDARs are presently limited to airborne craft.

REMOTE SENSING PLATFORMS AND SENSORS BEING CARRIED

Sensor Platforms

Sensors can be carried on space, air, terrestrial or water-borne platforms but it is beyond our remit to examine the latter two types. There are various major airborne or space platforms as follows:

1. *Balloons:* These may be free floating or anchored, and the former can be gas filled, hot air or propelled. They are now infrequently used, because they are slow, although there has been some discussion on bringing back into service the dirigible balloon.
2. *Helicopters:* They may be useful for the essential "ground truthing" work, i.e. collecting statistical data in more remote areas to verify images obtained from higher and faster platforms. They have rarely carried sensors directly.
3. *Space shuttles or Laboratories:* These are essentially manned space missions. They frequently carry experimental payloads which may require human testing or adjustment, or interactive participation with ground researchers. Any of the previously mentioned sensors can be utilised, and some of them, e.g. photography, with a great deal of flexibility.
4. *Airborne:* There are several sub-categories of this platform, according to flight altitude:
   (a) High altitude aircraft—these usually operate at over 8 000 meters allowing for photography at about a 1:100 000 scale, or the use of multi-spectral scanners and Radar systems.
   (b) Medium altitude aircraft—which operate at 3 000 to 8 000 meters and can take photographs at a 1:20 000 to 1:80 000 scale or carry multi-spectral scanners.
   (c) Light aircraft—which fly below 3 000 meters and can do aerial reconnaissance, take large-scale photographs and can supplement missing or uncertain satellite imagery.
5. *Satellites:* Since these are the platforms which provide the bulk of the remotely sensed data, and they are likely to be of increasing importance in the future. There have now been many hundreds of satellite launches, mostly by the U.S.A. and U.S.S.R., and the majority have been for military purposes. Recently several other countries

have launched their own satellites, either independently or in joint ventures. Initially, satellites were largely experimental, but they are now increasingly research or operational platforms, with most of them carrying a varied sensor payload. Their primary capability is to carry sensors which can monitor the entire earth surface on a periodic basis, sensing a large area during each revolution. They all operate at an altitude which is sufficient to escape from the earth's atmospheric drag but still remain within the dominant gravitational field, i.e. between 150 kms and 40 000 kms. Most satellites have been launched from rockets, whilst some have been launched aboard a shuttle spacecraft, from where they are unloaded into space. It has been recently demonstrated that the in-flight repair of satellite systems is now viable, though this is unlikely to be cost-effective for unmanned craft.

The advantages of satellites include the repetitive coverage of the earth's surface at various scales and at varying resolutions, with data being acquired on a routine and cost-effective basis. Often satellite-sensed data is the only information available for large tracts of ocean, mountain, desert or tropical forest areas. The disadvantages include their large capital costs, which include permanent monitoring and receiving stations, their relatively poor resolution for many environmental purposes and the fact that cloud cover remains a problem for many sensing devices.

Types of Environmental Satellites

It is convenient to classify environmental RS satellites into two major types, i.e. geostationary and near-polar orbiting.

Geostationary Satellites

These are satellite sensing systems which are boosted into a high geosynchronous orbit at approximately 35 900 kms above the equator, i.e. at this altitude the speed of the satellite can exactly match the speed of the earth's rotation. Because of this height, they have a limited number of uses, e.g. to transmit telecommunication signals or to get a broad view of the weather. The fact of remaining stationary means that they can achieve a high temporal resolution, but the great height means that spatial resolution is normally only in the range of 2 to 5 kms, according to wavelength. This type of satellite was first launched in 1966 and there are currently five geostationary satellites which each cover a different portion of the earth (Figure 8). They can image the earth's surface between latitudes 80N and 80S and they are able to image and transmit data on their whole viewable area every 30 minutes.

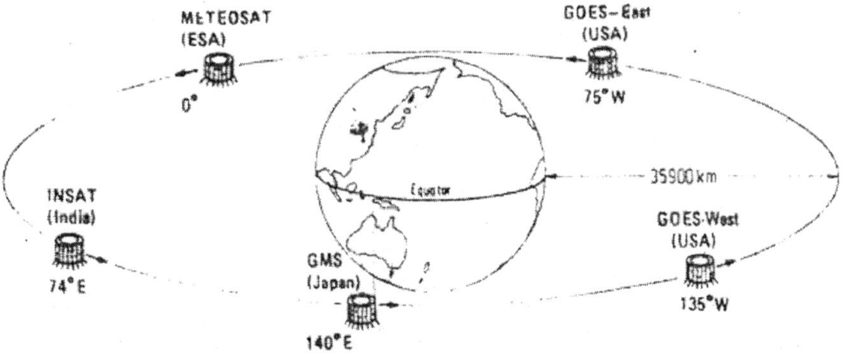

Figure 8: Positions and Names of the Five Geosynchronous Satellites Providing Meteorological Data.

Near-polar Orbiting Satellites

These satellites orbit the earth, with an inclination relative to the equator of nearly 90 degrees, i.e. so that their orbit nearly crosses the north and south poles (Figure 9). Their orbit height varies between about 270 kms and 1 600 kms and it is usually sun-synchronous—meaning that it crosses the equator at the same sun time each day. By having this type of orbit, the satellite visits any particular point above the earth at the same time, which is useful for the comparative analysis of multi-temporal data. One complete revolution of the earth takes about 95 to 115 minutes (depending on altitude), meaning that 12 to 16 revolutions are achieved each day. The exact inclination of the flight path will determine the time period between re-visits to any specific location, but it is commonly once every 16 to 20 days. These satellites have a working life expectancy of about four years.

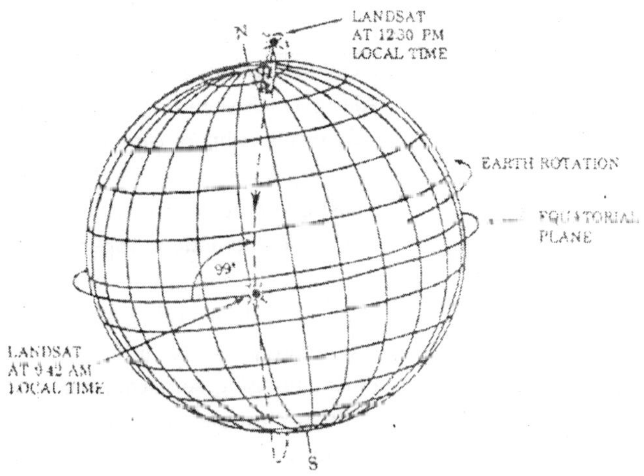

Figure 9: A Typical Orbital Track for a Polar Orbiting Satellite.

Major Operational Environmental Satellites

A number of environmental RS satellite systems and/or programmes have been in operation since the mid-1960s. Some of these programmes are ongoing, others have ceased. Table 2 attempts to list the main programmes.

Table 2: The Main Environmental Polar Orbiting Satellite Systems or Programmes

Satellite Programme	Country	Year of 1st Launch	Operational State	Sensors Carried
Tiros/NOAA 1st series.	U.S.A.	1970–1976	ceased	AVHRR; AVCS.
Landsat 1, 2, 3	U.S.A.	1972–1978	ceased	MSS; RBV.
Meteor	U.S.S.R.	1977	active	MSS; MRTVK.
Tiros-N/NOAA 2nd series.	U.S.A.	1978	active	AVHRR.
HCMM	U.S.A.	1978	ceased	HCMR.
Nimbus-7	U.S.A.	1978	ceased	CZCS; SMMR LIMBS.
Seasat-A	U.S.A.	1978	ceased	SMMR; RA; SASS; SAR; VIRR.
Landsat 4,5.	U.S.A.	1982–1982–1984	active	MSS; TM.
Kosmos	U.S.S.R.	1983	active	SLAR; MRIR; MRTVK. plus cameras
SPOT-1, 2.	France	1986	active	HRV.
IRS-1A	India	1988	active	LISS.
MOS-1a, b	Japan	1987–1990	active	MESSR; MSR; VTIR.
NROSS	U.S.A.	1989	active	Scatterometer
ERS-1	European Space Agency	1990	planned	AMI; SAR; ATSR-M; RA; Scatterometer.
Landsat 6.	U.S.A.	1992	planned	ETM:

AMI	Active microwave instrument.
ATSR-M	Along-Track scanning radiometer.
A VCS	Advanced vidicom camera.
A VHRR	Advanced very high resolution radiometer.
CZCS	Coastal zone colour scanner.
ETM	Enhanced thematic mapper.
HCMR	Heat capacity mapping radiometer.
HRV	High resolution visible instrument.
LIMBS	Profile temperature radiometer.
LISS	Linear imaging self scanning sensor.
MESSR	Multispectral electronic self-scanning radiometer.
MRIR	Medium resolution infrared radiometer.
MRTVK	Multispectral television system.
MSR	Microwave scanning radiometer.

Countd.

Countd.

MSS	Multispectral scanner.
RA	Radar altimeter.
RBV	Return beam vidicom system.
SAR	Synthetic aperture radar.
SASS	Seasat satellite scatterometer.
SLAR	Side looking airborne radar.
SMMR	Scanning multichannel microwave radiometer.
TM	Thematic mapper.
VIRR	Visible and infrared radiometer.
VTIR	Visible and thermal infrared radiometer.

Landsat 4 and 5

Landsat 4 and 5 were launched respectively in July, 1982 and March, 1984. They both have an angle of inclination of 98.3 degrees and an orbital time of 98.5 minutes. The satellites make 14to 15 revolutions per day and it takes 16 days before a revisit track is made. The Landsat distance between successive orbits. These satellites are a continuation of the original Earth Resources Technology Satellites (ERTS) programme, initiated in 1972, which later developed swathing pattern is illustrated in Figure 4.10, which also shows the width of swath and the into the Landsat series. Landsat 4 and 5 differ from the earlier Landsats by the introduction of the Thematic Mapper sensor and the exclusion of the Return-Beam Vidicom (Figure 4.11).

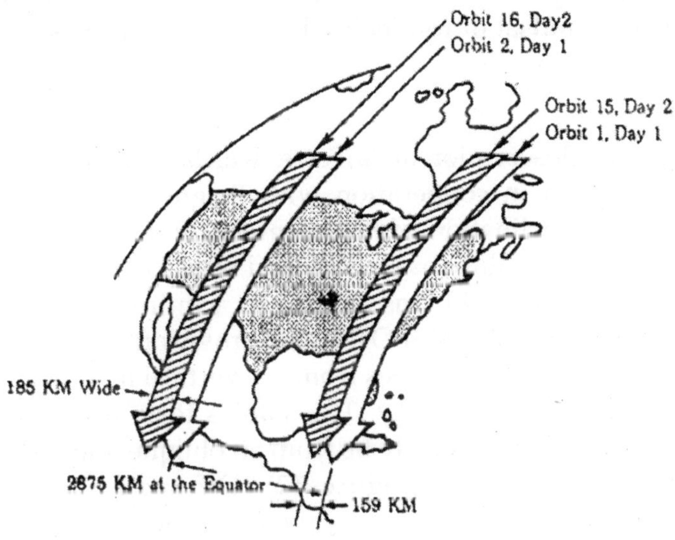

Figure 10: The Landsat Swathing Pattern and Successive Orbit Paths

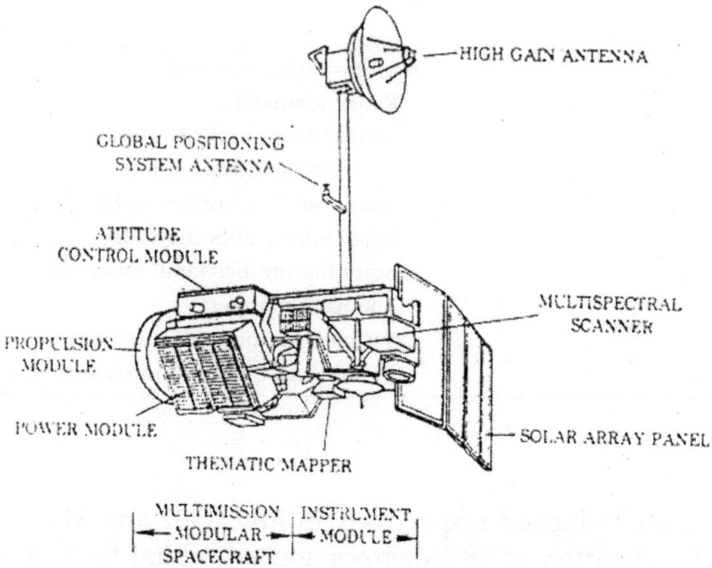

Figure 11: Configuration of Landsats 4 and 5

There are some 18 receiving stations around the world from where Landsat data is transmitted to NASA's Goddard Spaceflight Center. It is then passed to a commercial company (EOSAT) for processing and distribution. Both Landsat 4 and 5 have well outlived their life expectancy yet continue to operate because some of the on-board systems have been closed down to save power. There is likely to be a "data gap" when these two satellites finally cease to transmit, i.e. since lack of funding and technical problems will prevent Landsat 6 from being launched until the second quarter of 1992.

SPOT - 1

SPOT-1 (Système pour l'Observation de la Terre) was launched in February, 1986. It was constructed by the French in cooperation with Belgium and Sweden. It has a sun-synchronous orbit at an inclination of 98.7 degrees, and a revolution period of 101.4 minutes. It makes 14 or 15 revolutions per day and revisits the same track every 26 days. It's altitude varies from 820 to 840 kms and it can acquire images between 84 degrees North and South. A key feature of SPOT is the provision for off-nadir viewing, i.e. it can "look" side ways for up to 27 degrees from the vertical in either direction, extending the field of view by 475 kms each way. This allows for a much reduced revisit time, although images would then necessarily be from an oblique angle. This off-nadir facility is steerable from ground control (Figure 12). Off-nadir viewing also allows for stereoscopic viewing - pairs of images of a given scene can be recorded at different viewing angles during successive satellite passes in the vicinity of the scene concerned.

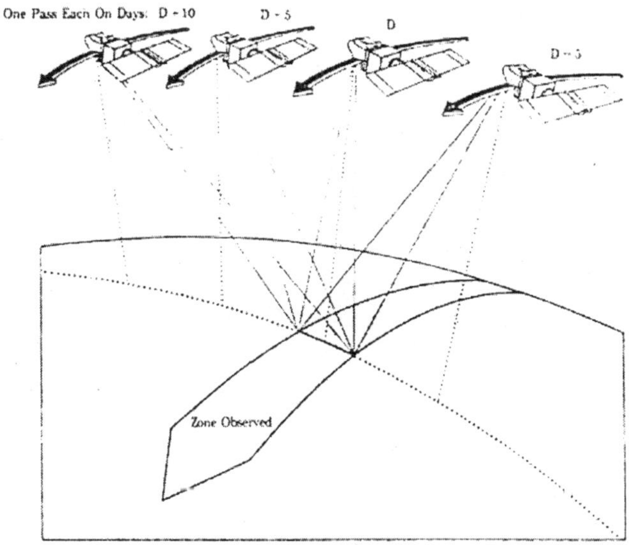

Figure 12: SPOT Off-Nadir Revisit Capabilities.

European data from SPOT is received at stations in Toulouse (France) and Kiruna (Sweden). On board recording of data is possible in areas beyond the range of ground receiving stations, for later transmission to Toulouse. Data can be received at stations over a maximum distance of 2 600 kms, and this means that transmissions can last for up to 800 seconds whilst passing a station. Stations are capable of receiving about 250 000 scenes a year. Each day, an observation sequence is loaded into the on-board computer from the Toulouse ground control station. Dissemination of SPOT data is via a commercial company "SPOT Image". SPOT-1 will be decommissioned in September, 1990. SPOT-2, with similar credentials to SPOT-1, was launched in January 1990 and two further satellites in the series are planned before 1998.

Marine Observation Satellite (MOS-1)

This was the first Japanese earth observation satellite - it was launched in February, 1987 and has a three-year scheduled life (Figure 13). It orbits at an altitude of 908.7 kms, having an inclination of 99.1 degrees and it makes 14 orbits per day. There is repeat coverage every 17 days and it takes 237 orbits to gain a total world wide coverage. MOS-1 does not carry tape recorders and thus ground stations are needed to acquire data which are out of range of the Japanese Earth Observation Centre. The European Space Agency has an agreement with the Japanese Space Development Agency to acquire, process and distribute MOS-1 products in Europe. MOS-1b, with identical credentials to MOS-1, was launched in February, 1990 and a further four satellites in the series are planned.

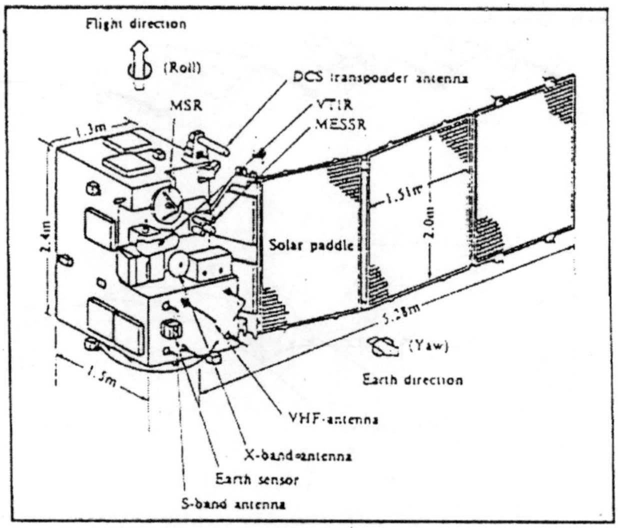

Figure 13: The Configuration of MOS-1

ESA Remote Sensing Satellite (ERS-1)

ERS-1 is to be the first of a series of satellites, in a programme to be operational in the 1990s, which have been planned by a consortium of countries making up the European Space Agency (ESA). Following its launch in 1991 ERS-1 will be placed in a sun-synchronous orbit and will give global coverage, including the polar regions. It will have an altitude of 777 kms, an inclination of 98.5 degrees and a revolution time of 100 minutes (Figure 14). ERS-2 is planned for launch in 1994. Real-time data will be relayed to stations at West Freugh (Scotland), Kiruna (Sweden), Fucino (Italy), Maspalomas (Canary Islands) and Prince Albert (Canada).

Kosmos

This Russian satellite series is a continuation of an older series, and it includes satellites launched for a variety of purposes. It is managed by the "PRIRODA" State Remote Sensing Centre. Soviet officials note that launches in the series occur once every two or three months. The satellites are placed in low (270 kms) orbits and the repeat time for full coverage is 22 days. Their life expectancy varies but it is very short. There is an absence of a network of ground receiving stations.

The Major Environmental Remote Sensors

Here we will be concerned to look first at the main sensors being carried by Landsats 4 and 5, SPOT 1, ERS-1, MOS-1 and Kosmos, including their usual applications. We will then describe two other sensors which have been particularly useful for marine or water based purposes.

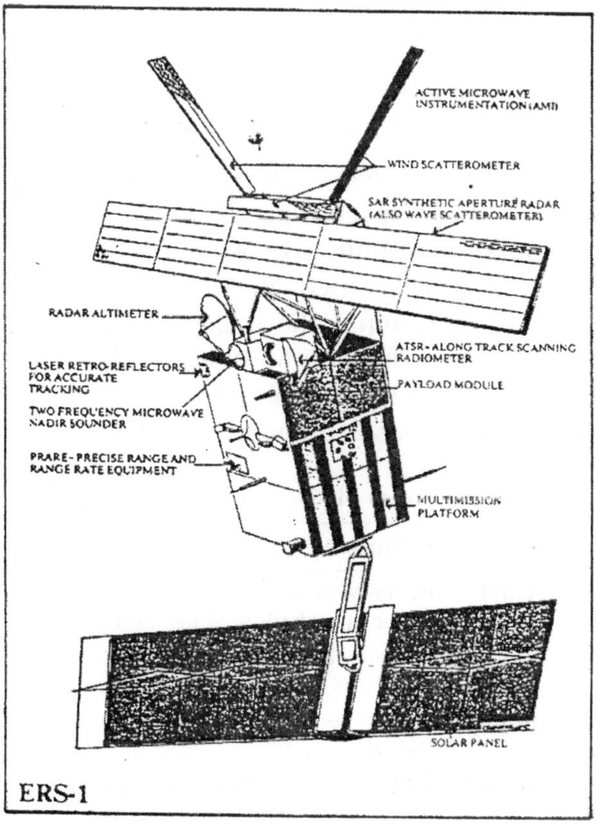

Figure 14: The Configuration of ERS-1

Landsat 4 and 5 sensors

Both Landsats carry similar sensors and there are two main types:

1. *Multi-Spectral Scanner*: This system is a line scanning device which records from four bands, two in the visible and two in the near infrared spectral bands. The system is comprised of a telescope, a mirror which reflects ground radiation onto a bank of 24 electro-optical sensors, band filters and a sampling system, an internal calibration system and various devices which ensure an orderly stream of digital data for each pixel and spectral band. It images six scan lines in each of the four spectral bands simultaneously giving a 24 scan-line total. Resolution, as delimited by pixel size, is 83 m and a six-bit quantisation gives a possible devices which ensure an orderly stream of digital data for each pixel and spectral band. It images six scan lines in each of the four spectral bands simultaneously giving a 24 scan-line total. Resolution, as delimited by pixel size, is 83 m and a six-bit quantisation gives a possible range of 64 intensity values. The individually scanned scenes of one MSS image covers approximately 185 x 185 kms and each overlaps its neighbour by about

10%. The original images have a scale of 1:3 369 000 and one frame encompasses 34 000 km². Data is recorded on magnetic tape for later transmission to receiving stations and data is available in digital or analogue (photographic) form. Table 3 shows details of the bands and possible applications of the imagery.

2. *Thematic Mapper:* This sensor collects, filters and detects radiation in a similar 185 km swath. It records in seven spectral bands which include medium and thermal infrared. It provides a spatial resolution of 30 meters, except on the thermal infrared band, where it is 120 meters. The high spectral resolution is achieved by sensitive detectors and an 8-bit quantisation in the analog-to-digital conversion process gives 256 grey levels. The Enhanced Thematic Mapper (ETM), to be launched on Landsat 6, will have an additional 15 meter resolution panchromatic band. Table 4 shows the potential application of this sensor by band and wavelength.

Table 3: Landsat MSS Bands and Applications

Band	Spectral Range	Features/Applications
4	500–600 nm Green band	Imagery from this band emphasises movement of sediment laden water and shallow water bodies, shoals and reefs, etc.
5	600–700 nm Red band	Imagery from this band emphasises cultural features, e.g. urban areas and roads, and sometimes bare soil surface colours.
6	700–800 nm Near infrared	This type of imagery emphasises vegetation and landforms in general.
7	800–1100 nm Second near infrared band	This imagery provides best penetration of atmospheric haze, and it emphasises vegetation and land-water boundaries.

Table 4: Bands and Applications of the Landsat Thematic Mapper

Band	Spectral Range (micrometers)	Resolution	Features/Applications
TM 1	0.45–0.52 visible blue- green	30 m	Bathymetry in less turbid waters, soil/vegetation differences, deciduous/coniferous differentiation, soil types.
TM 2	0.52–0.60 visible green	30 m	Indicator of growth rate/vegetation vigour, sedimentation concentration estimates, turbid water bathymetry.
TM3	0.63–0.69 visible red	30 m	Chlorophyll absorption/species differentiation, crop classification, tion, vegetation cover and density, geological applications.
TM 4	0.76–0.90 solar near infrared	30 m	Water body delineation, biomass and stress variations.
TM 5	1.55–1.75 solar mid infrared	30 m	Vegetation moisture/stress, minerals.
TM 6	10.4–12.5 emitted thermal	120 m	Surface apparent temperatures, urban versus land use separation, distinguishing burned areas from water bodies.
TM 7	2.08–2.35 solar mid infrared	30 m	Hydrothermally altered zones, mineral exploration, soil type discrimination.SPOT-1 sensors

SPOT carries two identical high resolution visible (HRV) scanners, each of which can function independently. They can each scan a strip measuring 60 x 80 kms along the flight line, this width varying with the viewing angle. Every 60 kms the SPOT data is cut to form a scene. The two sensors are designed to operate in either of two modes - panchromatic (black and white) or multi-spectral (colour) in the visible and near infrared spectral bands. The sensors are of the push-broom type. Each consists of a series of fixed linear arrays made up of electronic detectors known as CCDs (charged-coupled devices). Image data are collected by successively measuring the current generated by each detector along the array. In panchromatic mode each individual detector corresponds to one pixel and measures the reflectance of a 10 meter ground resolution cell. In multispectral mode the detectors are paired and thus measure a 20 meter pixel across the track. By doubling the time taken to obtain each sample, the along track measurement of the cell also becomes 20 meters.

We have already mentioned the nadir, off-nadir and stereoscopic viewing capabilities of the system. The main applications of the stereoscopic imagery are in photogrammetry, for cartographic purposes and photo-interpretation for geological, geomorphological and hydrological studies. SPOT"s other application are largely for land-use studies, the assessment of renewable resources and aiding in mineral and oil exploration. The high resolution makes possible the compilation of topographic maps (at scales of 1:100,000), with the contour interval as little as 20 meters, thematic mapping of between 1:25,00 and 1:50 000 plus the direct compilation of digital terrain models. Table 5 gives the main characteristics of the HRV.

Table 5: Characteristics of the HRVs on SPOT-1

Characteristics of the HRV instrument	Multispectral mode	Panchromatic mode
3 Spectral bands:green	0.50–0.59?m	
red	0.61–0.69?m	
near-infrared	0.81–0.91?m	
or		
1 Broad spectral band:		0.50–0.75?m
Instrument field of view	4.13°	4.13°
Ground sampling interval (nadir viewing)	20m×20m	10m×10m
Number of pixels per line	3000	6000
Ground swath width (nadir viewing)	60km	60km
Pixel coding format	3×8 bits	6 bits DPCM (1)
Image data bit rate	25M bits/s	25 M bits/s

ERS-1 Sensors

The main sensors on board will be:

1. Active Microwave Instrument(AMI)combining the functions of a Synthetic Aperture Radar (SAR), a Wave Scatterometer and a Wind Scatterometer. The AMI will

measure wind fields and wave spectra and obtain all-weather images. It will provide for a spatial resolution of 30 meters and have a swath of 99 kms.

2. Radar Altimeter (RA) to measure significant wave height and provide measurements over ice and major oceanic currents.
3. Along-Track Scanning Radiometer (ATSR-M) to determine sea surface temperatures and measure atmospheric water vapour.
4. Precise Range and Range Rate Experiment (PRARE) for accurate satellite ranging and to allow ionospheric error correction.The programme objectives include the global observation of: waves; sea state ;ocean currents; ocean waves; sea surface temperatures; sea ice; ice sheet dynamics, and also to provide imaging of land by SAR. These objectives will help with shipping, fisheries and offshore activities.

MOS-1 Sensors

The details of the three sensors on board MOS-1a and 1b are given in Table 6. MOS is intended to establish fundamental technologies for earth observation satellites, primarily by observing oceanic phenomena such as ocean colour and temperature. The satellite observations are also expected to be of value to agriculture, forestry, fishery and environmental preservation.

Table 6 Bands, Features and Objectives of MOS-1 Sensors

	MESSR	VTIR		MSR
Objective	sea surface colour, vegetation, land use, etc.	suspended sediment	stratospheric water vapour,Earth and sea surface temperatures	water vapour content, liquid water content, ice,snowetc.
Observation Wavelength(?)	0.51–0.59 0.61–0.69 0.73–0.80 0.80–1.10	0.5–0.7	6–7 10.5–11.5 11.5–12.5	
Observation Frequency (GHz)				23.8±0.2, 31.4±0.25
Beam Width				1.89±0.19, 1.31±0.13
Integration Time (m sec)				10 & 47, 10 & 47
IFOV(km)	0.05	0.9	2.7	32, 23
Swath Width(km)	100(each optic)	1500		317

Kosmos Sensors

Satellites in the series have carried different sensors and combinations of sensors. Of most interest for environmental observations are their cameras. The KFA-1000 provides for a ground resolution as small as 5 meters, giving it a great advantage over SPOT or

Landsat. Another advantage is the frequency of cover. 98% of all PRIRODA's RS imagery is obtained from a combination of the three cameras carried on Kosmos satellites. After imaging in space, the exposed film is soft-landed to earth for processing.

Other Sensors of Interest to Aquaculture and Inland Fisheries

There have been a number of other satellite sensors launched in the past two decades which have been of potential value to aquaculture and inland fisheries. We will briefly describe two of these:

1. *The Heat Capacity Mapping Radiometer (HCMR):* This sensor was launched as part of the Heat Capacity Mapping Mission in April 1978 and operated until September 1980. The sensor was a two channel scanning radiometer operating in the visible and near infrared band and the thermal infrared band. The main objectives of interest were:
   (a) The mapping of natural and man-made thermal effluents.
   (b) The detection of thermal gradients in water bodies.
   (c) The mapping and monitoring of snow fields for water run-off prediction.
   (d) The monitoring of marine oil pollution.
      (i) Many of the products obtained from this mission are still available.
2. *The Coastal Zone Colour Scanner (CZCS):* This was launched aboard Nimbus-7 in October 1978 and was operational until late 1984. The CZCS was a multi-spectral line scanner, optimised for use over water. It collected quantitative information on ocean colour, suspended sediments, chlorophyll concentrations, pollutants and temperature from the upper few meters of water. Both photographic and digital data are still available. Table 7 sets out the bands and measurements of CZCS.

Table 7 Bands and Measurements of the CZCS Sensor

Wavelength (pm)		Spatial Resolution	Swath Width	Measurments
Band 1	0.43–0.45	800m	1800km	chlorophyll absorption
Band 2	0.51–0.53	800m	1800km	chlorophyll distribution
Band 3	0.54–0.56	800m	1800km	gelbstoffe(yellow substance)
Band 4	0.66–0.68	800m	1800km	chlorophyll concentration
Band 5	0.70–0.80	800m	1800km	surface vegetation
Band 6	10.50–12.50	800km	1800	surface temperature, diffuse attention co-efficient

DATA PROCESSING OF REMOTELY SENSED IMAGERY

The electronic images which have been captured by RS devices are either transmitted directly to earth or are stored on on-board recorders for later transmission. This represents the initial stage in a complex information flow which is depicted in Estes . The data scanned are retained in the form of pixel values, with each value representing the amount of radiation(the spectral reflectance) within a given band-width received by the scanner from the area of the earth's surface covered by the pixel. Pixel values are digitally coded by a certain number of bits, i.e. Landsat and SPOT use 8-bit codings which give a range of 256 possible values, and the values for any one pixel will change according to particular spectral bands being recorded. The area covered by a pixel (the resolution) is a function of the height of the sensor, the focal length of the lens or focusing system, the wavelength of the radiation and other inherent characteristics of the sensor itself. Each pixel will be allocated a co-ordinate in agrid referencing system.

The pixel values are transmitted to earth (downlinked) as a stream of binary numbers. To reconstitute the images, ground based computers decode the binary data, allocating the appropriate colour tone to each pixel value. The images can then be displayed on a monitor or in some print-out form. At the initial stage they will be monochrome and in a pre-processed state.

Image Pre-Processing

The two functions involved here, radiometric and geometric corrections, will be essential if meaningful output is to be obtained. They are essential because there are a number of factors inherent in the RS system which contribute to the images being distorted in some way. These factors include:

1. Changes in the attitude, velocity and altitude of the sensing platform.
2. The forward motion of the platform causes scan skew.
3. The scanners (in Landsat) do not have a constant scan velocity.
4. The area covered by one pixel will have its shape distorted when viewing at an oblique angle.
5. The geometry of the images is affected by the earth's rotation, its curvature and atmospheric refraction.
6. Radiometry is affected by the sensor, e.g. sensor "noise" and poor calibration between detectors, by the atmosphere, e.g. presence of aerosols and scattering effect, and by the scene itself, e.g. effect of relief on reflection and type of reflection of the object.

Radiometric Correction: Detector sensitivity will slowly change over time, making some detectors more or less sensitive to radiance than its neighbours. This results in images which have a "banding" or striped pattern which needs correcting. This effect can occur in both mechanical or push broom scanners, as can "pixel drop" when individual pixel radiance is not recorded. Radiometric distortions (atmospheric attenuation) is also a problem as radiance is altered by the atmosphere through which it passes. This is especially a problem over water when there is a lot of atmospheric water vapour, i.e. radiance reaching the detector may be 20% from the water, 80% from the atmosphere. There are various correction methods, some of which are described in Butler et al.

Geometric Correction: This will involve several levels of pre-processing. The data has first to be corrected for earth curvature, earth rotation and satellite attitude errors. After this the image may still contain geometric distortions, with the center of the scene located to an accuracy of only a few kilometers. To improve this, a sufficient number of ground control points, which are readily identifiable on the image and on a map, are selected for calculations of a least-square fit, and the results are then used to adjust the image to the map co-ordinates. Maps of different projections, e.g. Mercator, Peter's Conformal Lambert, etc., can be used.

Image Display and Enhancement

There are a large number of functions which can be further performed on the digital data in order to meet specific requirements. Enhancements are designed to aid the analysis of images and pictures so that the maximum use can be made of them, e.g. the use of colour rather than grey tones means that more discrimination can be applied.

1. *Contrast stretching:* Sensors are designed to record a full range of values from a very strong signal (usually displayed as white) to no signal (black), with in-between some 254 grey tone steps. Few scenes require such a dynamic range and data values for any scene tend to occupy less than half of the available range. The values can be changed to occupy the full scale by using a stretch programme that can be applied to give a linear, uniform distribution, gaussian or other arbitrary stretch.

2. *Edge enhancement:* This technique can be visualised as using the computer to recognise subtle but sudden steps in the grey scale, where the sudden step in local grey scale is continuous over a regional variation in background digital values. Its effect is to increase the grey scale of the step so that its edge is more clearly displayed. It is particularly useful for the detection of linear elements, either oceanic (warm or cold water fronts) or man-made (canals, waterways, etc.).

3. *Density slicing:* This involves delineating the different ground features by colour coding, i.e. areas having similar reflectance values are assigned to one colour. Up to eight "slices" might be taken through the range of pixel values which would produce an image showing eight key features in eight colour ranges. The computer can calculate when to best define a threshold to separate individual colours and it can also calculate the area covered by the different features (colours or colour tones). It can be particularly useful in water depth analyses. A simpler way of using colour is to select three spectral bands and feed the data from each through a different colour gun on the image processors monitor. This results in either a natural-colour or false-colour composite. False-colour indicates that colours have been assigned to

information sensed in bands which are beyond the visible range, i.e. they simulate radiation which cannot be seen.

4. *Spatial Filtering:* This term includes a family of operations which use values for groups of pixels to spatially filter areas to produce images with different effects such as smoothing, or line enhancement. For example, "low pass" filters assign to each pixel of the radiance values to be suppressed. On the other hand "high pass" filters can enhance differences of reflectance values in the image.
5. *Principle component analysis (Data compression):* This analysis compresses multi-spectral data sets as an aid to image classification and pattern recognition. The normally highly correlated spectral bands are transformed into a new set of variables or components that are uncorrelated. The new components represent more efficiently the variance of the data set. Using Landsat MSS data the first two principle components generally account for over 95% of the original variance, although lower order components may include useful image detail.
6. *Image classification.* Spectral classification techniques can be grouped into supervised and unsupervised classifications.
7. *Multi-temporal analysis (Change detection):* Data sets collected at different dates (multi-date) or at different times (multi-temporal) of the same day can be registered and compared pixel by pixel. The pixel value of each pair can then be ratioed or subtracted. Other procedures can also be used, such as the statistical procedures of Principle Component Analyses, to compose the multiple data sets. The resulting digital numbers will highlight differences between the images. This is of particular use in studies of crop state, seasonal vegetation anomalies, rivers, thermal changes, coastal processes, etc.
8. *Digital mosaics:* A display application of interest is the "patchworking" of images from the same sensor to form mosaics covering large areas. This poses radiometric and geometric problems since the selected scenes may have been imaged at different seasons of the year, i.e. when colour balances are different. Mosaics are of value in studying regional features of land cover, geology, natural resources and in cartography.

THE POTENTIAL APPLICATIONS OF REMOTE SENSING TO AQUACULTURE AND INLAND FISHERIES

Water Quality

A large number of variables many influence water quality, and thus potential fish productivity. Though it is not yet possible to use RS techniques to detect variations in most of these qualitative factors, it is possible to gain a number of inferences as to water quality, e.g. by looking at:

1. Areas affected by acid rain.

2. Land use and crop yields.
3. Effluent discharges and other pollutants.
4. Suspended and cohesive sediment concentrations.
5. Salinity.
6. Phytoplankton concentrations.

Because some of these variables exhibit similar radiometric values, it has proved necessary to establish various empirical models to show the relationship between water radiance values and different environmental parameters. Even when these models have been verified they are not always globally valid. Further problems with assessing water quality via RS exist, such as the fact that the spatial resolution is insufficiently detailed for most river assessment and the fact that most radiance values will only refer to near surface water. Additionally, since many water qualitative parameters undergo frequent changes, the monitoring frequency from satellite-derived data is insufficient, and indeed even "real time" images may not be real enough in the case of some pollutants.

1. *Acid rain:* It is recognised that this is an increasing problem in several major world regions, and that areas where vegetation is showing signs of damage will also exhibit water having adverse acidic conditions. Using Landsat TM imagery obtained over southern Germany, in the wavelengths of 0.75 to 1.3 um, damage assessment could be determined since affected trees gave significant spectral changes. Information derived from reference data on tree species, age, degree of damage, etc., is being used to train and test image classification and for the interpretation of results.

2. *Land use and crop yields*: Trolier and Philipson have shown the feasibility of again using TM scenes, only this time by visual analysis, to determine classes of land use (from 22 different classes) which could have a major effect on the quality and quantity of watershed run-off. This proved quite successful using TM Bands 3, 4 and 5 or a composite of them. Clearly, different land use types will give clues as to the likely degree of pollution to be expected, though it would be wise to seek verifiable data to back up that gained from remotely sensed imagery. Larger urban and industrial areas can readily prove to aggravate water quality as can areas having intensive farming with high crop yields. Also areas which have exploitive activities such as mining, quarrying or forestry are likely to be qualitatively affected. Most land uses which lack human intervention are easily detected and would exhibit better quality water.

3. *Effluent discharges and other pollutants:* RS techniques are practicable for the detection of high volume effluents whose spectral characteristics are markedly different from the surrounding water, e.g. effluents which are highly coloured or at elevated temperatures. Examples of pollutants monitored from satellites include sewage sludge dumped at sea, sewage outfalls, titanium dioxide and oil discharges, spills or blowouts.

4. *Suspended and cohesive sediment concentrations:* Radiation in the visible part of the spectrum penetrates significantly below the water surface where it is absorbed and scattered. The presence of suspended particles causes much "backscatter", giving the water a cloudy appearance. The Landsat MSS, the SPOT HRV and the CZCS have all allowed false-colour composite images to be prepared which clearly highlight sediment concentration and plumes. A colour slice enhancement from a single band (of the MSS) has been used to quantitatively map sediment concentrations, though reliable calibration requires the collection of water samples at the time of the satellite overpass. Sediment concentrations can be used as natural tracers to identify water movement and frontal boundaries. Cheney and Rabanal report much of the work done on the measurement of suspended sediment concentrations from satellite data, warning that there is an identity overlap with information on chlorophyll content. Ritchie and Cooper report the success of the satellite monitoring and estimation, using Landsat MSS data, of sediment concentrations in small inland water bodies.
5. *Salinity*: Both Butler et al and Cordell and Nolte report that salinity level measurements from satellite data are not yet recorded in the literature, but that they are certainly feasible with the use of microwave sensors, i.e. changes in salinity cause significant changes in the emissive brightness temperature of water for frequencies less than 5 GHz. The measurement of this brightness should be adequate for mapping the near-surface spread of freshwater at a river mouth or estuary. Although salt wedges at some depth could not be plotted, brightness measurements could be a great help in site selection for many forms of aquaculture, both in saline conditions and at the mariculture/fresh water interface.
6. *Phytoplankton concentrations:* This water qualitative factor is certainly the best documented, possibly because detection is both very important and it provides dramatic large-scale imagery which is not usually visible, e.g. occurrences of significant algal blooms in the northern Adriatic and the Scandinavian Skagerrak area.

Water Temperature

A number of sensors, working in the thermal spectral range of 2 to 18 um, are capable of detecting surface water temperatures. The main sensors used to data have been the MSS, CZCS, AVHRR, SMMR and the TM. Sea surface temperature (SST) sensors operate from both geostationary and orbiting satellites. Since data obtained from the GOES geostationary satellite necessarily has a large spatial resolution (8 kms), it is only useful for sea, large lake or oceanic (macro-scale) areas, though on the plus side data can be obtained at 30 minute intervals. For smaller water bodies, the AVHRR instruments aboard the NOAA satellite series is most frequently used. It offers a spatial resolution of about 1.1 kms and a thermal resolution of 0.2 degrees C. Many SST algorithms have been developed by NOAA and tested in a wide variety of situations, and the validity of satellite-derived SST data is now widely recognised for bodies of water of various sizes. For smaller bodies of water Band 6 of the TM on-board Landsats 4 and 5, has provided satellite thermal imagery at a resolution of 0.12 kms.

Economic Production Functions Related to Distance Factors or Population Distributions

A large number of production function are concerned with spatial distributions of a "man-made" social and economic environment, e.g. such factors as distance to the various markets, size of markets and the availability of various economic inputs. Since oblique remotely sensed images are analogous to maps, i.e. showing 2D distributions of spatially related objects at different scales, then they can be used in virtually the same way as maps. Once the scale has been determined, RS images can give direct information concerning distances, but they can only give inferred information on market size, input availability and land costs. SPOT HRV data is ideal for mapping, and updating maps, at scales of >1:25,000. Most human developments of importance, e.g. roads and urban areas, can be discriminated and Iisaka and Hegedus have shown how, using appropriate models, reliable population estimates (inferring market size) can be made from Landsat MSS images. Once the digitising of existing base maps has been accomplished for any given area, then updating can be continuously undertaken as more recent geometrically corrected imagery is acquired, and as automatic feature recognition techniques are applied and improved.

Climatic Factors

Many factors relating to weather and climate can, and are, being obtained and compiled from satellite RS imagery, and this data has been of help in compiling both short and long-term meteorological information. However, this data is of little direct use to prospective fish producers since, for the optimising of production locations, the enterepreneur must be concerned with not only average long-term conditions but also known climatic extremes. This information would already be in existence for virtually all areas, i.e. at the only realistic scales which are very small. Short-term forecasting could be of value in indicating that a potentially hazardous event was imminent - one for which precautionary measures might be advisable, e.g. exceptionally strong winds or very heavy rain. Models are now being developed where by future runs of particular weather conditions can be reliably inferred from present weather forecasts, e.g. of temperatures, wind directions or air pressure, though for the immediate future these are only likely to be successful for oceanic areas where there is spatially extensive uniformity of physical conditions.

Soils

Soil types can be inferred from RS images both directly and indirectly. Since natural vegetation types correlate strongly with soils, then the various vegetation categories will give a clue to the suitability of soils for pond construction. Soil type may also correlate with altitude, various climatic factors and with slope, and certain soil types will relate to areas of intensive crop production. Any indirect RS determination of soils must be carefully calibrated and substantiated by ground truthing. Several studies have shown that Landsat TM imagery is the most suitable for recording the very varied spectral reflectance characteristics of different soils, i.e. spectral signatures will vary with the amount of organic matter, mineral content, iron oxides present, moisture content, surface roughness as well as the percentage balance of sand, silt and clay. Post-soil classification evaluations of the predicted results show that soils can be categorised with almost total accuracy, and the area of each soil type can also be accurately estimated. Some confusion may be apparent between bare soils and areas of very low vegetation cover.

Relief

The ability of the SPOT HRV instrument to perform off-nadir viewing means that it is possible to obtain stereo-pairs of images, which in turn can be used for the plotting of contours, i.e. with the use of an Ortoster instrument. 50-metre contours can be plotted at a scale of 1:50,000. This is generally sufficient to determine whether slopes in a particular area would prove to be disadvantageous to aquaculture site location.

Existence of Shelter for Cage Culture

The existence of shelter for sea cage culture will be a function of several factors, i.e. strength and direction of prevailing wind, length of fetch and the height of surrounding land. We have seen that relief is now able to be determined—this will give an indication as to potential shelter. The Seasat satellite, which was operational for a short time in 1978, proved the possibility of gaining very accurate measurements of wave height, i.e. with the RA instrument an accuracy to within 10 cms could be obtained, and another instrument on board, the SASS, measured surface wind speeds and direction in all weather conditions. From these instrument readings fairly accurate charts could be drawn up, which gave long-term average or extreme wave or wind conditions. The prospective ERS-1 satellite, will soon be able to continue to provide this data.

Bathymetry

Again knowledge of this function is important to cage culture. There are several methods of determining water depth in areas where it is relatively shallow (<20 m). Where water visibility is clear, the Landsat MSS instrument (using Band 4) can detect reflectance from the sea bed. At different depths, a different degree of brightness will be measured and bathymetric contours can then be drawn based on lines of equal brightness. In more turbid areas, then images using Band 4 reflectance captured on the Landsat TM sensor, can be preprocessed and then density-sliced to discriminate between water and sandbanks. The actual height of sub-surface sandbanks can be determined by using multiple images at different tidal states, and, by using images captured over a period of time, sandbank dynamics can be calculated. There are some other production functions which can be inferred from satellite imagery, e.g. monitoring of sea ice; monitoring of vegetation levels or biomass estimates in inland waters and the existence of coastal flora or habitat types such as mangroves, but space precludes an analysis of these. During the next decade it is anticipated that sensors with higher resolutions will become available, and not only will this improve upon the imagery available but it will also allow satellite platforms to be used to gain data on production functions which presently rely on airborne sensing.

ACQUISITION OF REMOTE SENSING INFORMATION

RS is a field of study which has not only generated a vast amount of new and usable data at an exponential rate, but it has spawned information about itself at a similar rate. Cordell and Nolte, in a study which was aimed at testing the feasibility of using RS to help in aquacultural location, noted—"The quantity of information (RS) to choose from is so large that it will tend to confuse and discourage customers who are unfamiliar with remotely sensed information.". It has been our experience too, that those who are interested in RS will start to delve and very soon they will suffer from a surfeit of information.

Remotely Sensed Product Types

The wide range of RS products will have been created via the enactment of a large number of highly technological processes. A number of satellite systems are producing different forms of output which spawn different products, but since the vast majority of output is presently, and easily, available from two main satellite series, landsat and SPOT. The quality of data output is quite variable and therefore customers may stipulate minimum quality acceptance levels, maximum cloud cover acceptance and preferred times of year for an image. It is possible to obtain "quick-look" microfiche or photographic images before purchase to assess image quality. To order products the user should be familiar with the spatial referencing system appertaining to the satellite series.

Digital Data or Computer Compatible Tapes (CCTs)

For SPOT images CCTs can be recorded at either 6250 or 1600 bits per inch (bpi). Given that the total volume of a SPOT scene (which is normally 60 × 60 kms but 60 × 80 kms for the maximum oblique viewing angle) is 27 to 100 Mbytes, then a single 6250 bpi tape can accommodate the entire scene. Two or three 1600 bpi tapes would be needed per scene depending on the scene data volume. Alphanumeric characters are coded by default in ASCII or optionally in EBCDIC, and files are organised according to "Band Interleaved by Line" mode or "Band Sequential" mode. Imagery on CCTs will have been pre-processed to one of six levels depending on customer choice. CCTs showing Landsat imagery contain similar characteristics to SPOT for data volume and format. However, the levels of pre-processing will be different and more varied since data is derived from three sensors (MSS, RBV and TM) and there are three basic scene sizes available—full scene, quarter scene or a user defined map scene. A single Landsat full scene will cover about 9 times the area of a SPOT scene, though at a reduced resolution.

Photographic Products

Both SPOT and Landsat images are available in black and white or colour formats, in film or as prints, and they may be at varying sizes which correspond to the scale of the image (see Tables 8 and 9). For both Landsat and SPOT special photographic products are available, e.g. for colour composites various specific bands from each Landsat sensor are used by default, but any combination of bands can be requested. SPOT special photographic products include various mosaics, colour composites, linear combinations of spectral bands, panchromatic photographs at 1:25 000 scale, stereoscopic images, "shift along track" or extended frame scenes, digital elevation models and orthophotos. Landsat also provide a similar range of special products. The range of photographic products available from the Kosmos series (via PRIRODA) is shown in Table 10.

Table 8: Availability of Landsat Photographic Products

Sensor	Nominal Image Size	Film	Paper	Scale
RBV Landsat 1,2 and	18.5cm (7.3ins)	X	X	1:1,000,000
MSS Landsat 1 to 5.	37.1cm (14.6ins)		X	1:500,000
	74.2cm (29.2ins)		X	1:250,000
RBV Landsat 3.	18.5cm (7.3ins)	X	X	1:500,000
	37.1cm (14.6ins)		X	1:250,000
	74.2cm (29.2ins)		X	1:125,000
TM Landsat 4 and 5.	18.5cm (7.3ins)	X	X	1:1,000,000
	37.1cm (14.6ins)		X	1:500,000
	74.2cm (29.2ins)		X	1:250,000

Table 9: Availability of SPOT Photographic Products

Scene	Nominal Image Size	Film	Paper	Scale	Processing Level
Whole	241cm×241cm	X		1:400,000	1A,1B,S1,2,S2
	300cm×350cm	X		1:400,000	2,S2
	482cm×482cm	X	X	1:200,000	1A,1B,S1
	964cm×964cm		X	1:100,000	1A,1B,S1
Quarter	241cm×241cm	X		1:200,000	1A,1B,S1
	482cm×482cm	X	X	1:100,000	1A,1B,S1
	964cm×964cm		X	1:50 000	1A,1B,S1

Table 10: Kosmos RS Products Available With Prices (U.S.$)

Survey camera and image type	Film positive	Film negative	Film negative in set with film positive	Contact print on paper	Paper print, 2X enlargement	Paper print, 4–5X enlargement
KFA-1000						
Scale and image format	1:270,000 30×30 cm	1:270,000 30×30 cm	1:270,000 30×30 cm	1:270,000 30×30cm	1:130,000 4 parts, 30×30 cm	1:50,000 25 parts, 40×40 cm
Black and white	$1,180	$1,260	$1,360	$440	$660	$1,150
Color	$1,560	$1,650	$1,770	$550	$820	$1,400
MK-4						
Average scalea and image format	1:900,000 18×18cm	1:900,000 18×18cm	1:900,000 18×18cm	1:900,000 18×18cm	1:500,000 36×36 cm	1:250,000 4 parts, 40×40 cm
Black and white, 1 band	$1,050	$1,110	$1,190	$210	$300	$440
Black and white, 3 bands	$2,520	$2,660	$2,860	$420	$600	$880
Color	$1,950	$2,070	$2,250	$700	$1,020	$1,600
Color synthesized	$3,220	$3,470	$3,800	$750	$1,070	$1,600
KATE-200						
Scale and image format	1:1,350,000 18×18 cm	1:1,350,000 18×18 cm	1:1,350,000 18×18 cm	1:1,350,000 18×18 cm	1:700,000 36×36 cm	1:350,000 4 parts, 40×40 cm
Black and white, 1 band	$230	$250	$280	$100	$140	$220
Black and white, 3 bands	$550	$580	$630	$210	$310	$470
Color	$450	$480	$510	$200	$280	$420
Color synthesized	$690	$750	$790	$235	$300	$450

Product Availability

RS products are available via a number of companies and organisations whose status varies from government level through to private commercial companies. With incrasing volumes of data available, with rapidly rising costs in the space industry and with changing government priorities, a distinct trend towards a commercial outlook in the provision of data is obvious. Three major commercial companies are the primary source of all SPOT, Landsat and Kosmos imagery:

1. *SPOT Image:* This company, based in France, was formed in 1982 to handle all commercial distribution of SPOT imagery, to promote the system and to generate and process data. They have established a network of distribution points in over 20

countries. All SPOT scenes are archived and then referenced in a computerised data base called the "SPOT Image Catalogue" which can be accessed via telephone, telex and switched-packet international networks at any time. They also offer extensive guidance and support facilities. They have increased their business considerably since 1986, though they are never likely to cover satellite development costs.

2. *EOSAT:* The Earth Observation Satellite Corporation was formed in 1985 as a joint venture by the Hughes Aircraft and the R.C.A. Corporations. It aims to promote future satellites in the Landsat and Omnistar series, develop new sensor systems, to promote and market Landsat imagery and to decrease government costs in RS. It presently receives a government subsidy of $19 million annually to cover Landsat 4 and 5 costs. It operates a network of receiving stations around the world and has its base at Lanham in Maryland. It holds over two million Landsat Images in a worldwide accessible digital archive and it also provides extensive user support facilities.

3. *Soyuzkarta:* This company was set up by PRIRODA in 1987, both to help achieve economic accountability of the Russian RS industry and to accelerate efforts to increase sales of photographic products on a world-wide basis. A vast geographic range of imagery is available, though it is all in photographic form and it would therefore need to be digitised before using in a GIS. Over 100 long-term contracts have already been made with "western" countries to supply imagery which uis supplied through a company called Central Trading Systems. There is a desire by PRIRODA to increase contacts with foreign firms to help economically and to acquire better technology and expertise.

These three companies have various agreements with other companies and organisations to supply RS imagery. These include:

1. *EURIMAGE:* This is a consortium of companies, formed in 1986, which have the right to market Landsat data on a commercial basis throughout Europe. They are based in Rome and work under the auspices of the European Space Agency (ESA).

2. *EARTHNET:* This is the ESA distribution network, based in Rome, which handles the European data from SEASAT, the Nimbus 7 CZCS, the HCMM, and from the NOAA and Meteosat satellites.

3. *EROS:* The Earth Resources Observation Systems Data Center is part of the U.S. Geological Survey, being located at Sioux Falls, South Dakota. It is responsible for archiving, reproducing and distributing the Landsat data held by the U.S.G.S., through current Landsat data is distributed through EOSAT. It therefore has a large selection of older images.

4. *NESDIS:* The National Environmental Satellite, Data and Information Service is part of NOAA in the U.S., and it provides a large range of RS products associated with meteorology, the ocean and the environment.

5. *NCIC:* The National Cartographic Information Center, in the U.S., also provides a range of map, photographic and satellite imagery.
6. *NRSC:* The National Remote Sensing Centre is situated at Farnborough in the U.K., where it is the main source for all satellite imagery. It provides an extensive user support system.

All these distributors have various forms of on-line, microfiche or microfilm data retrieval systems, many of which are accessible at other points or via computer telecommunication links. Computer listings of the data held are usually available. Although some of these companies and organisations purport to supply data obtained from Russian, Japanese or Indian satellite systems, in reality this may be hard to procure. Spatial coverage of satellite imagery is now very comprehensive. Landsat TM and MSS data covers the whole world except a small exclusion zone over western China and the central Soviet Union. Having said this, there would be wide variations in the number of scenes available for any particular location. Figure 15 illustrates the amount of this variation over the comparatively small area of England, Wales and Scotland. The variations are mostly due to cloud cover problems, but they may also be a function of the satellite orbit return frequency and the degree of swath overlap, plus the systems performance of the spacecraft, e.g. systems performance of Landsat was very poor during 1983 and 1984 before Landsat 5 was in full operation.

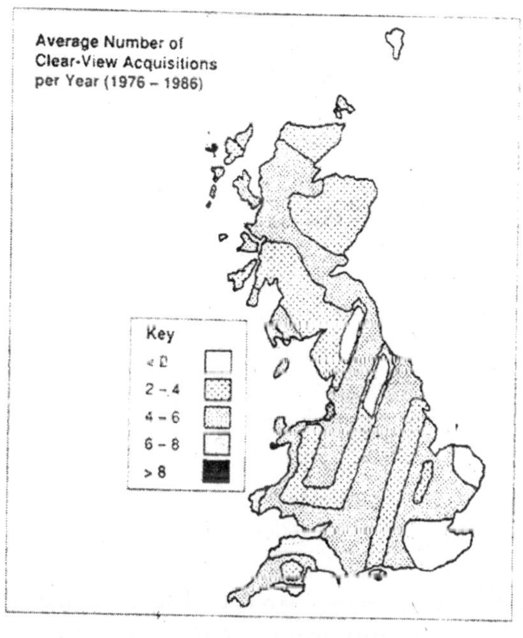

Figure 15: Variations in Landsat Data Acquisition.

The time taken to acquire imagery is usually two to three weeks, though by paying a high premium turn-around time can be as short as 48 hours. Prices for satellite imagery vary greatly according to products; an indication of 1989 Landsat prices is given in Table 11. SPOT prices are generally higher per km² of scene, especially for high resolution products and for those having undergone maximum pre-processing. Because of the desire to earn foreign exchange, prices for high resolution Soviet imagery are cheaper than either Landsat or SPOT, even after essential digitising costs have been met.

Table 11: Selected Landsat Imagery Prices for 1990

TM DIGITAL DECODED:		
Full scene	185×170km	$4900
Quarter scene	100×100km	3600
Map sheet	55.5×91.2km	2500
TM DIGITAL MOVABLE SCENE (User defined):		
	100 × 100km	$3200
	50 × 100km(N-S × E-W)	2400
	100 × 50km(N-S × E-W)	2400
TM COLOUR PRODUCTS (185 × 170 km):		
Positive transparency	1:1 000 000	$1800
Paper	1:1 000 000	$1300
Paper	1:500 000	1400
Paper	1:250 000	$1500
TM BLACK AND WHITE PRODUCTS (185 × 170 km):		
Positive transparency	1:1 000 000	$500
Paper	1:1 000 000	550
Paper	1:500 000	600
Paper	1:250 000	700
MSS DIGIOTAL PRODUCTS:		
	185 × 170km	$1000
MSS COLOUR COMPOSITES (185 × 170km):		
Positive transparency	1:1 000 000	$600
Paper	1:1 000 000	550
Paper	1:500 000	700
Paper	1:250 000	1000
MSS BLACK AND WHITE PRODUCTS (185 × 170km):		
Positive transparency	1:1 000 000	$155
Paper	1:1 000 000	95
Paper	1:500 000	190
Paper	1:250 000	290

It must be remembered that the costs shown in Table 12 are just for the data. Hardware and software costs for data manipulation are high and the costs of analyzing the data can also be high, especially if consultancy fees are necessary or if training costs are considered. It is likely that the commercial companies are charging prices which are too high for the research/scientific community, e.g. sufficient Landsat CCT images to cover the entire land area of the earth would be >$50 million. Having said this, if the costs of using RS imagery are calculated per km^2, then they may be considered to be very reasonable. In a recent study to test the cost-effectiveness of RS techniques for aquaculture site assessment, for the Lake Patzcuaro area in Mexico, it was found that costs for digital MSS imagery plus necessary ground truthing were about 33% lower than for a conventional ground-based survey, and 38% lower than for a survey which combined aerial photography with ground-based methods.

Guidance and Support Facilities

Support given by the Major Companies and Organisations

The major companies and organisations involved in RS offered a degree of guidance and support. The range of guidance and support facilities will vary as a reflection of the primary aims of the company or organisation. To exemplify will examine support given at three levels:

1. *Commercial companies:* Since their primary aim must be to maximise financial returns, their range of support will be limited. It will mostly fall under the broad notion of "marketing". Apart from selling the main CCT and photographic products they produce helpful material which maximises the likelihood of selling more primary RS data. Examples of this would be :
   (a) Computer listings of images available.
   (b) Directories or manuals containing product information about sources and uses of material.
   (c) Floppy discs containing imagery data or software programmes for data manipulation.
   (d) Posters and brochures aimed at company and product promotion.
   (e) Computerised data base catalogues or other product retrieval aids.
2. *National remote sensing organisations:* These will be government organisations whose raison d'être is to promote an "industry" in which much government investment may have been made. A much broader approach to promotion can be made because financial considerations are not paramount. As well as providing the support that commercial companies do, and acting as agents to distribute the products from the commercial companies, they may also:
   (a) Issue regular newsletters concerning the whole "industry".
   (b) Offer consultancy services to specialist users.

(c) Carry out research aimed at broadening the use of RS.

(d) Offer facilities such as film writing, image analysis or image processing at their premises.

(e) Distribute a wide range of promotional materials such as posters, fact-sheets, slide sets, videos, product catalogues, etc.

(f) Run various educational or training programmes.

(g) Operate libraries with archive or browse facilities.

(h) Operate travelling exhibitions geared to a range of venues.

3. *The FAO Remote Sensing Centre (FAORSC):* Operating on a global scale, the organisation's objective is to use space technology to help bring about self-reliance of developing countries. They offer many of the support facilities listed for commercial companies and for national organisations, but dealings are usually at a governmental level and a list of the type of support offered reflects this:

(a) Advising countries on the suitability, availability, etc. of RS data, and on methods, techniques and equipment.

(b) Conducting practical training courses at national or regional levels.

(c) Developing RS centers and national RS networks.

(d) Undertaking pilot studies to demonstrate the utility of RS.

(e) Promoting RS technology and applications through technical publications and technical support.

(f) Identifying and developing RS projects to help with resource assessment, management and exploitation.

(g) Providing geographic data on the availability of RS imagery.

(h) Promoting environmental monitoring by the use of RS.

(i) Providing early warning systems on food security concerns.

(j) Promoting international technical cooperation in space applications between member nations and international organisations.

Other Sources of Guidance and Support

World-wide there are now a large number of learned and professional societies which may be related to RS in various ways. These societies propagate information and support by holding conferences and symposia, by issuing special publications, technical papers and conference proceedings and by organising courses, workshops and regular meetings.

World-wide there are also numerous university courses which are either completely devoted to RS or which RS forms a significant part of. For a select bibliography, which covers RS at a general or introductory level, we would recommend: Curran, Lo, Mather, Hyatt and Drury. Cordell and Nolte provides a detailed practical summary, directed towards aquaculture, of how to acquire RS information. Finally, Table 12 offers a list of some available journals whose content is mostly devoted to RS.

Table 12: Journals Which are Mostly Devoted to Remote Sensing

(a) Arianespace.
(b) Atlantic Canada Remote Sensing Newsletter.
(c) Aviation Week and Space Technology.
(d) EARSEL News.
(e) Earth Observation Quarterly.
(f) E.S.A. Journal.
(g) European Space Report
(h) Geocarto International.
(i) Geoscience and Remote Sensing.
(j) ISPRS Journal of Photogrammetry and Remote Sensing.
(k) ITC Journal.
(l) International Archives of Photogrammetry and Remote Sensing.
(m) Int. Journal of Aerial and Space Imaging, Remote Sensing and Integrated Geographical Systems.
(n) Int. Journal of Remote Sensing.
(o) Journal of Geophysical Research.
(p) Journal of IEEE Geosciences and Remote Sensing.
(q) La Lettre du CNES.
(r) Mapping Sciences and Remote Sensing.
(s) Photogrammetric Engineering and Remote Sensing.
(t) Remote Sensing in Canada.
(u) Remote Sensing Letters.
(v) Remote Sensing of Environment.
(w) Remote Sensing Society Newsletter.
(x) Space News.
(y) Space Policy.

PROBLEMS OF UTILISING REMOTE SENSING METHODOLOGIES

Inherent in the introduction of any comparatively new, large-scale and complex technology there will be problems-indeed the very fact of making progress is achieved by overcoming problems. So in some senses, it could be argued, we should not be concerned with them. However, since some authors have argued that RS has failed to live up to its proclaimed potential, and within the RS community it is freely acknowledged that its capabilities may have been oversold, then it is only reasonable to highlight a few of the major problems.

Variable Amount of Accessible Data

Data accessibility is variable spatially, i.e. within a comparatively short distance it may vary by a factor of five or more. This variability is caused mostly by cloud cover and temporal variations in the functioning of sensors. Add here that certain climatic zones pose particular problems to RS, i.e. especially the humid tropics where cloud cover is frequently high, plus west-facing coasts in the mid to high latitudes where onshore prevailing winds cause high cloud cover, and those areas which have seasonal rainfall due to movements in the inter-tropical convergence zone. Some areas may lack data for other reasons, e.g. politically sensitive areas, areas beyond the receiving range of some satellite transmitters and high latitude areas.

Necessity for Ground Truthing

It is necessary to verify the imagery, i.e. by matching up various known classes of land cover with pixel values displayed on images for the given area. This is necessary since pixel values for any specific point will vary from day to day according to a number of extraneous factors. This means that RS use has to constantly consider this inconvenience in terms of cost, problems of accessibility and the necessity of ground truthing at approximately the same time as the satellite overpass.

Accessibility to Suitable Hardware

It is very clear that RS produces vast amounts of digital data and that whatever form this is used in expensive hardware will be required to process it. The more that the use of the data can be maximised, the more expensive will be the necessary hardware. The spatial areas and the types of programmes where RS techniques are likely to prove most valuable are also those which may be at a physical distance from any suitable hardware. Although data is beginning to be available in cheaper forms, e.g. floppy disks, and hardware will undoubtedly proliferate and fall in price, cost and accessibility will continue to make many potential RS using programmes non-viable.

Spatial Resolution

For many purposes airborne sensing still offers the highest resolution and the lowest price, and until resolutions get down to about the five meter size then RS imagery will continue to have its limitations. Although the technology exists to secure better resolution levels, there are political, cost and legal barriers to be overcome before it is generally available. Though the present SPOT resolution of 10 meters is valuable, there are many facets of the human landscape, especially in developing countries, which are not easily determined or cannot be detected at all. Though Wright has noted, with a degree of truth, that the present crude pixel size could be an advantage given that the data volume problem would be greatly exacerbated with a smaller pixel size, RS is likely to have far greater user acceptance and project applicability with an improvement in resolution.

Commercialisation of RS Products

The increasing trend to commercialise RS output throws up a number of quite significant problems. Clearly products sold will need to have very stringent copyright regulations which themselves will need careful interpretation, e.g. who has the right to purchase individual scenes and can anyone else then purchase exactly the same scene? Agreements which must be carefully set up on a world-wide basis to operate product outlets or agencies, and this has led to a new and complex field of "space law". Allied to these commercial problems are those of where to fix prices-there is the divergence in the need to recover large costs yet the need to sell data, often to organisations in the less developed world who can ill afford expensive imagery.

Long Term Planning and Payload Uncertainty

Most potential users of RS require assured access to imagery over a long temporal period, even if it is simply to justify the necessary hardware and software expenses for one project. A major problem that the industry faces is, what has been called elsewhere, "planning blight". Virtually all satellite launches have suffered from uncertainty in scheduling and payloads, with most programmes being significantly behind schedule. For example, ERS-1 will have suffered at least a three-year launch delay, and Landsat 6 was supposed to have carried a SEA-WIFS sensor but this has been abandoned. Delays have been caused by a lack of funding, technological problems, major disasters such as Challenger, etc. It is difficult to sell a technology which has such an unreliable track record, and any short or long-term planning is at the best vague. Whilst there are many other minor problems, e.g. delays in data acquisition, lack of trained personnel, etc., the application of RS to aquaculture and inland fisheries does not in fact suffer problems as much as the application would to many other enterprises or fields. This is because, except for some of the water qualitative parameters, production functions controlling fish production do not require real-time or regular repeats of imagery, and given that RS can only really help location optimising at the macro scale, then resolution is also not a problem.

REFERENCES

Elangovan,K (2006)"GIS: Fundamentals, Applications and Implementations", New India Publishing Agency, New Delhi"208 pp.

Harvey, Francis(2008) *A Primer of GIS, Fundamental geographic and cartographic concepts.* The Guilford Press, 31 pp.

Heywood, I., Cornelius, S., and Carver, S. (2006) *An Introduction to Geographical Information Systems.* Prentice Hall. 3rd edition.

Longley, P.A., Goodchild, M.F., Maguire, D.J. and Rhind, D.W. (2005) *Geographic Information Systems and Science.* Chichester: Wiley. 2nd edition.

Maguire, D.J., Goodchild M.F., Rhind D.W. (1997) "Geographic Information Systems: principles, and applications" Longman Scientific and Technical, Harlow.

Ott, T. and Swiaczny, F. (2001) *Time-integrative GIS. Management and analysis of spatio-temporal data,* Berlin / Heidelberg / New York: Springer.

9

GPS Applications in Crop Production

Global Positioning System (GPS) provides opportunities for agricultural producers to manage their land and crop production more precisely. Common names for general GPS applications in farming and ranching are precision agriculture, site-specific farming and prescription farming. GPS applications in farming include equipment guidance such as sprayers, fertiliser applicators and tillage implements to reduce excess overlap and skips. They can also be used to precisely locate soil-sampling sites, map weed, disease and insect infestations in fields and apply variable rate crop inputs, and in conjunction with yield monitors, record crop yields in fields.

WAYS OF GPS USED IN FARMING

Equipment Guidance Systems

Lightbar-guided and automated steering systems assist in maintaining precise swath-to-swath widths. Guidance systems identify an imaginary A-B starting line, curve or circle for parallel swathing using GPS positions and a control module. The module takes into account the swath width of the implement and then uses GPS to guide machines along parallel, curved, or circular evenly spaced swaths. Guidance systems include a display module that uses audible tones or lights as directional indicators for the operator. The guidance system allows the operator to monitor the lightbar to maintain the desired distance from the previous swath. Guidance systems require two principle components: (1) a light bar or screen, which is essentially an electronic display showing a machine's deviation from the intended position and (2) a GPS receiver for locating the position. This receiver must be designed for this purpose and it must operate at a higher frequency (position calculations are usually 5 to 10 times per second) than a GPS receiver needed to record positions for a yield monitor. GPS receivers designed for guidance can be used in conjunction with a yield monitor or for other positioning equipment. Automated steering systems integrate GPS guidance capabilities into the vehicle steering system.

Automated steering frees the operator from steering the equipment except at corners and at the end of fields.Guidance systems base prices are approximately $3000, including the GPS receiver and a readout unit. Systems that steer the vehicle will be higher.

Yield Monitoring Systems

Yield monitoring systems typically utilise a mass flow sensor for continuous measuring of the harvested weight of the crop. The sensor is normally located at the top of the clean grain elevator. As the grain is conveyed into the grain tank, it strikes the sensor and the amount of force applied to the sensor represents the recorded yield. While this is happening, the grain is being tested for moisture to adjust the yield value accordingly. At the same time, a sensor is detecting header position to determine whether or not yield data should be recorded. Header width is normally entered manually in to the monitor and a GPS, radar or a wheel rotation sensor is used to determine travel speed. The data is displayed on a monitor located in the combine cab and is stored on a computer card for transfer to an office computer for analysis. Yield monitors require regular calibration to account for varying conditions, crops and test weights. Yield monitoring systems cost approximately $3000 to $4000, not including the cost of the GPS unit.

Field Mapping with GPS and GIS

GPS technology is used to locate and map regions of fields such as high weed, disease and pest infestations. Rocks, potholes, power lines, tree rows, broken drain tile, poorly drained regions and other landmarks can also be recorded for future reference. GPS is used to locate and map soil-sampling locations allowing growers to develop contour maps showing fertility variations throughout fields. The various datasets are added as map layers in geographic information system (GIS) computer programs. GIS programs are used to analyze and correlate information between GIS layers.

Precision Crop Input Applications

GPS technology is used to vary crop inputs throughout a field based on GIS maps or real-time sensing of crop conditions. Variable rate technology requires a GPS receiver, a computer controller, and a regulated drive mechanism mounted on the applicator. Crop input equipment such as planters or chemical applicators can be equipped to vary one or several products simultaneously. Variable rate technology is used to varying fertiliser, seed, herbicide, fungicide and insecticide rates, and for adjusting irrigation applications. The cost all of the components necessary for variable rate application of several products is approximately $15,000, not including the cost of the GPS receiver. Technology capable of varying just one product costs approximately $4,000.

HOW DOES THE GPS SYSTEM WORK?

GPS System

Precisely locating positions on Earth is not a new phenomenon. Navigators, sailors, explorers and surveyors have done this for centuries as they traveled about the world. Most maps and globes display longitude and latitude or some other coordinate projection information. Points on Earth are given unique addresses on maps using specific coordinate systems. Agriculturists commonly use either a geographic system of latitude and longitude measured in degrees or a Universal Transverse Mercator coordinate system that locates positions in meters measured from a specific point. The GPS system uses measured distances to the precisely located GPS satellites to locate positions on Earth. Radio receivers in GPS units monitor radio signals broadcast from the GPS satellites. A GPS position is determined by simultaneously measuring the distance to at least three satellites. The distance to a satellite is measured by the time it takes a radio signal to travel from the satellite to the GPS receiver. Computers in GPS units use information from the radio signals, including broadcast time and unique satellite information, to calculate positions. Information from at least four satellites is needed to calculate elevation. Signal reception from more satellites increases position accuracy. The global positioning system includes a constellation of 24 systematically arranged satellites orbiting the earth in 6 orbital planes with 4 satellites in each plane. The satellite orbits are approximately 12,500 miles above the earth. The constellation is arranged to guarantee radio reception from at least 4 satellites from any location anytime, anywhere on Earth. GPS receivers normally receive signals from 8 to 9 satellites in location without obstructions like buildings or trees.

GPS Errors

The quality of GPS units and operational errors associated with the GPS system determine the accuracy of GPS-located positions. There are several sources of GPS errors. GPS radio signals can "bounce off" objects such as buildings and trees prior to acquisition by the GPS receiver resulting in lower accuracy. This is called *multi-path error*. (Figure 1).

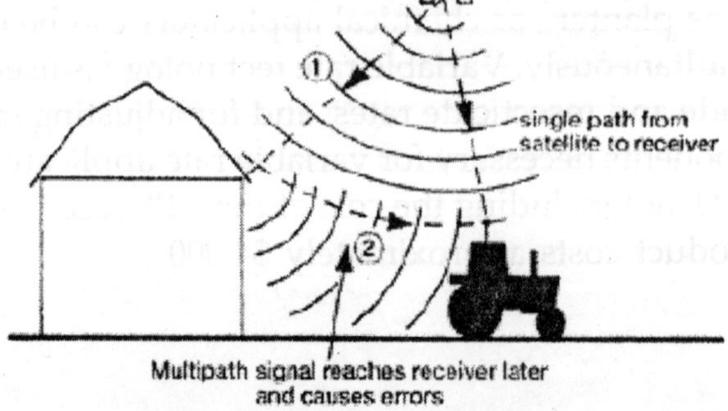

Figure 1. Multipath signal errors

The satellites use very accurate atomic clocks to generate the timing data that is received by the GPS receivers. However, even small *errors in timing* from clocks in the satellites and GPS units cause errors in GPS positions.

Signal delay errors can be caused by atmospheric interference such as electrically charged particles in the ionosphere. A layer of water vapor located below the troposphere can also alter the speed of travel of the radio signals.

Errors from GPS satellites' orbit and location are also significant. Pressures from solar radiation and gravitational forces of the sun and moon can alter satellite locations.

GPS receiver quality also effects GPS accuracy. More costly GPS units generally provide more accurate GPS positions than less expensive units.

WHAT IS DIFFERENTIAL CORRECTION?

Types of Correction

Differential global positioning systems (DGPS) reduce GPS errors and provide more accurate and reliable readings. Differential correction uses a radio signal broadcast from known locations on Earth. These Earth-based stations receive radio signals from the GPS satellites and determine the error from their known positions. The error is calculated and transmitted the GPS receivers. The US government and several commercial companies provide differential correction GPS services.

The *US Coast Guard* provides a free differential correction *Beacon Signal*. The Coast Guard signal is an AM radio signal that is broadcast from several locations, and travels as a "ground wave" over the earth's terrain. Each station has a radial coverage of approximately 300 miles. As the distance increases, the accuracy of the signal decreases. This system was originally designed for use on US navigable waterways but is being expanded to provide coverage over much of the US.

Several commercial geo-stationary satellite differential correction services are available for subscription charges. These systems use known Earth-based stations that receive GPS satellite signals and determine the amount of error. The corrected signal is broadcast to a geo-stationary satellite, which in turn broadcasts the corrected signal to the subscribed GPS receivers. (Figure 2)

The *Wide Area Augmentation System (WAAS) differential correction* system uses a network of 25 ground-based reference stations. The US Federal Aviation Administration operates the WAAS differential correction service to provide accurate GPS positions for commercial aircraft. The reference stations relay GPS-determined locations to a master station. The master station calculates a correction factor that is transmitted to geo-stationary satellites. WAAS-enabled GPS units receive the corrected signal from the WAAS geo-stationary satellite. The WAAS system uses two geo-stationary satellites located over the east and west coast of the United States.

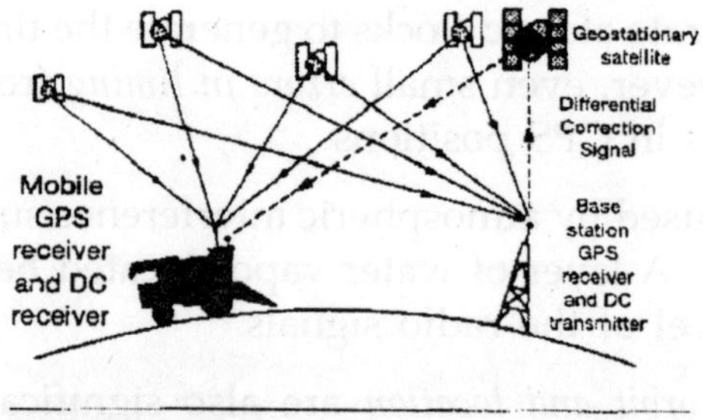

Satellite-based differential correction systems use geo-stationary satellites to relay the differential information from base station receivers to moving receivers

Figure 2. Geo-stationary DGPS representation

Real-time Kinematic Differential Correction

A fourth type of differential GPS correction, commonly called real-time kinematic GPS (RTK GPS), provides GPS position accuracy to within 1 centimeter. RTK GPS requires a separate base station located within approximately 5 miles of the mobile GPS units. The RTK base station is a known location equipped with a GPS unit. The base station GPS location is corrected to its known location, and the correction factor is transmitted to the mobile GPS units by FM radio signals. The accuracy of RTK GPS results from the close proximity and of the base correction station.

Levels of Accuracy – Receiver Types – Costs

GPS units for precision agriculture applications require sub-meter accuracy, must incorporate differential correction, and are priced from approximately $1,500. Handheld GPS units without differential correction locate positions within about a 30-foot radial area and range in price from less than $100 to many times that amount. Some handheld GPS units are available with differential correction. RTK GPS systems cost several times as much as common sub-meter accuracy GPS unit used in agriculture.

REFERENCES

Ess, D., & Morgan, M., (1997). The Precision-Farming Guide for Agriculturists. (1st ed.). Moline, IL.

Stombaugh, T., Shearer, S., Fulton, J., (2002). GPS Simplified. 2002 University of Kentucky Ext. Rept. PA-5.

Yeung, A. K. W., & Lo, C. P., (2002). Concepts and Techniques of Geographic Information Systems. Upper Saddle River, NJ.

10

ICTs in Precision Agriculture

The recent advances in information and telecommunication technologies have allowed farmers to acquire vast amounts of site-specific data for their fields, with the ultimate aim being to reduce uncertainty in decision-making. Precision Agriculture (PA) or site-specific crop management can be defined as the management of spatial and temporal variability at a sub-field level to improve economic returns and reduce environmental impact. Within the concept of PA the main activities are data collection and processing and variable rate applications of inputs. The tools available consist of a wide range of techniques and technologies from information and communication technology as well as sensor and application technologies, farm management and economics. PA is intrinsically information intensive, and farmers face many difficulties in efficiently managing the enormous amount of data they collect. They may lack sufficient time or are reluctant to invest the time needed to analyze the data and interpret the information. Additionally, the economic benefits of PA practices have not been proven yet and especially applications such as variable rate nitrogen application and patch spraying.

The first applications of PA around the world started in the early nineties but the initial adoption started at the end of the nineties. Yield monitors connected to a GPS receiver was the first real attempt to conduct site-specific management on their fields and continued with variable rate applications of mainly lime, fertiliser and recently chemicals. Soil sampling has also been also a very important PA application; however the sample size varies among the farmers depending mainly on the soil analysis cost. Sensing devices, such as Electromagnetic Induction (EC) measuring soil structure and water content and the Hydro-Nitrogen sensor, which senses the chlorophyll and automatically adjust the fertiliser dozes, have been very promising, but farmers are still skeptical on their reliability and effectiveness. Finally, remote sensing has recently attracted the focus on many research projects, while commercial companies and advisors are using aerial photos or satellite pictures to relate them with yield potential, nutrients' deficiencies and stresses. When commercial yield mapping started, it was expected that parts of a field would constantly yield well, while other parts would produce poor results.

This was due to the assumption that permanent soil characteristics would always behave in the same way each year. Therefore, many researchers have devoted their time to develop trend maps, both spatial and temporal over a number of years to be able to show these trends in the fields. While these expectations, the spatial trends that were expected to become more stable over time, Blackmore *et al.* proved that spatial trends over 6 years in four fields in the UK became less pronounced than the variability found in individual years. This was a significant reversal of existing wisdom and has profound implications on how spatial and temporal variability should be managed. Each year, individual yield maps often show significant spatial variability but they appear to cancel out each other over time. A major implication of these findings is that the treatment of fields based entirely on historical yield trends can no longer be supported. The corollary of this is that management should now concentrate more on managing the variability within one year. In this matter, the design of DSS should incorporate such factors. The lack of DSS in PA as the main impediment in adopting PA has been also supported by a wide range of researchers.

GLOBAL ADOPTION OF PRECISION AGRICULTURE

The global adoption of yield monitors has been predominated in North America, Europe and Australia but countries like Argentina, Brazil and some East Asian countries have also adopted PA practices. Nowadays, the adoption rate is in stationary state, mainly since yield increases are not well documented to cover the cost of equipment, together with the agricultural commodity prices that are continually falling. Have seen similar trends with other high-tech technologies. It took for instance about 10 years to adopt the Internet and mobile phones among the large majority of consumers. In this respect see a similar or even slower adoption pattern and a reduction in production costs of PA technologies in line with these technologies. Technical enthusiasts were the first to adopt PA practices. In the USA, about 90% of the yield monitors in the world are operating there. The main crops that use PA practices are corn, soybeans and recently cotton. Griffin et al. have referred to specific numbers of yield monitors, soil 3 sampling, remote sensing and variable rate applications in the USA and around the world. In the USA, it is estimated that in 2003, there should be around 45.000 combines equipped with yield monitor where about 46% of corn, 36% of soybeans and 15% of wheat had harvested with a combine equipped with the yield monitor.

In Europe, the adoption rates are far smaller than in the USA. It is assumed that about 400 Danish, 400 British, 300 Swedish and 200 German farmers have adopted yield monitors by the year 2000. A general trend is that larger farmers, with more than 300 ha, tend to be the first to invest in the new technology, whereas small farmers are more reluctant to invest in GPS equipment. A nation wide survey in America concludes that the adoption of PA technologies has been related to farm size and large farmers are the first to adopt. Compared with other farm technologies such as GMO's (Genetical

Modified Crops), the adoption of PA practices has been relatively slow. The main reason is that PA is time demanding approach, while GMO crops are time-saving, which is highly appreciated by farm managers. On top of that, the techniques and the management skills to use the GM-crops was already established at the launch of the seeds, while in PA, the agronomic and economical interpretation as well as the data analysis was still in its infancy, when the first practices were used by farmers. It seems however, that the adoption rate has been a little faster in North America compared to Europe. A reason for that is likely to be larger farm sizes and specialisation in certain cash crops. Whereas the European crop producers traditionally have been smaller in terms of farm size and less specialised compared with their American colleagues.

The adoption of yield monitors and especially satellite pictures have been relatively popular in the US compared with Europe—probably because of large areas with cash crops and less cloudy days in the landlocked areas. Variable treatment with fertilisers and weed detection seems however to be well adopted in Europe. In the US, the application of yield monitors without GPS positioning is a rather common practice in most states, where it is estimated that only one third of the yield monitors are connected to a GPS receiver.

INNOVATION AND DIFFUSION ASPECTS OF PRECISION AGRICULTURE

The development from the first idea to practical application of a new technology can be defined as a process in three steps: Invention, innovation and diffusion. Invention usually relates to the first idea of using a new technique. In PA, the invention was the first research trials to map the site-specific yield with yield meters on the field according to longitude and latitude using the GPS-system. A technical innovation is defined as the first commercial application or production of a new process or product. Innovation has also been defined by Nielson and Winter as a change of decision rules to fit with the surrounding requirements. An innovation includes all activities that create technological change. It may depend on several activities such as research and development, product design, process engineering, distribution and marketing. Diffusion however, is related to the wide spread adoption of an innovation among potential users. What causes a successful diffusion and how can we expect a general adoption among the users?

The primarily parameters in the process of diffusion are the population of potential adopters and their process of decision making. Moreover, the interacting flow of information between the adopters and manufactures has an impact as well. It also seems evident that the size of the potential users and their economic capabilities have an impact on diffusion. Moreover, studies have shown 4 that the complexity of the innovation might influence the diffusion and adoption speed and the general impact on productivity among adopters. S. Davies has developed a model that characterises two different types of innovation. The first model is the A group which illustrates the labour productivity for technologies that are fairly simple. The learning effects of this simple technology is initially very high but after some time the productivity gains from that technology will be limited implying that the curve falls and stabilises at a given level.

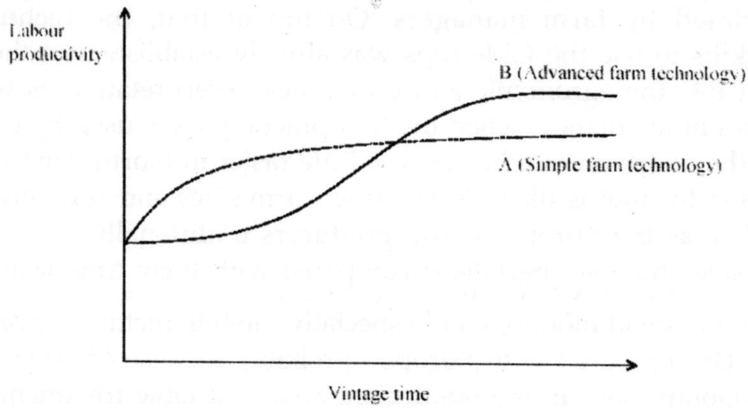

Figure 1: Labour productivity for group A and B innovations Pedersen 2003 based on S. Davies.

The other group of technologies (group B) be can be characterised as fairly complex. It requires some time to be familiar with the technology and the productivity gains from that technology will take place at a much later stage compared with the technologies in group A. Methods of sitespecific application is likely to follow the process as described in group B. The concept of PA is highly technical and requires skills that go beyond technical knowledge. It requires an understanding of the interrelated impact that various data and information have on yields, crop quality and nutrient leaching. The adoption and diffusion of various techniques for practising PA is therefore dependent on the provision of advice and technical as well as biological management guidelines. Griffin et al., 2004 argued that it is important to understand farmers' reasons for non-adoption first and then try to help the farmers address their questions to make them more confident adopting the technologies to use. A number of detailed studies refer to the profitability of PA and reasons for non-adoption.

ICT ADOPTION ASPECTS IN PRECISION AGRICULTURE

Addresses the findings of six mail surveys, focus groups and personal interviews with key stakeholders of PA in the UK, USA and Denmark over the last six years. The information related to ICT adoption in PA was extracted and is presented here. Additionally, a global search on ICT issues in PA has given the state-of-the-art in this domain. The issues presented here are innovation and diffusion of PA, demographics, PA software, data ownership, data handling, data interpretation, Internet and e-mail use and information sources to invest and practice PA.

Demographics of Precision Agriculture

The practitioners of PA tend to belong to a younger generation and they cultivate larger areas than the average farmer. The average age of the Danish respondents was 43 years old and 46 for the American respondents. In Denmark, the average age of farmers in 2000 was 52 years old and in the USA in 2002 was 55,3. The average cultivated area per farm in Denmark and the US Corn Belt was 422 ha and 790 ha respectively. The average cultivated area of the survey group was considerably higher than the average farm holding in both Denmark and the US. In comparison, an average farm area in Denmark is about 50 ha and the average farm holding in the USA is about 178 ha, while in the State of Indiana the average is about 102 ha.

Precision Agriculture Software

There are several types of PA software. The most common are software to generate maps (e.g. yield, soil); software to filtering collected data; software to generate variable rate applications maps (e.g. for fertiliser, lime, chemicals); software to overlay different maps; and software to provide advanced geostatistical features. There are a few companies that operate world-wide and provide integrated software packages from generating all different types of maps, having statistical analysis tools and also record keeping. The machinery companies that provide yield meters also offer software to generate yield maps and fertiliser companies provide software to generate variable rate applications maps. Finally, universities as well as regional companies provide a range of software for filtering of data, advanced statistical tools and record keeping incorporating PA data. Some of the packages are very complicated for farmers to use and they are fairly expensive, while some others are considerably simpler and cheaper with fewer options.

According to interviews with farmers about their perceptions on the available PA software, they believe that the software has been improved tremendously over the last 5 years. The packages are more user-friendly and have many options for the farmer to use. However, there are still problems related to data transfer between farmers, and between farmer, co-op and consultant. To overlay maps, mainly soil and yield maps, is also a difficult task so far. One farmer mentioned that the PA software has to be as fast and easy as the "Microsoft Windows programmes". Other farmers mentioned, that just to keep up with new products was a significant task. The majority of the PA farmers who do yield mapping own the software themselves, while only a small number of farmers who practice variable rate applications own the software. About 65% of 6 the respondents who did yield mapping own the software in Denmark and 90% in the US Corn Belt. Some 48% of the Danish respondents who have used VR fertiliser applications own the software for VR applications and 38% of the American respondents. Danish respondents own less software for VR applications than hardware (48% own the software and 71% the hardware), while the American respondents own more software than hardware (38% own the software, 27% the hardware). In the Danish case, this may be explained by the fact that local advisors produce the application maps for farmers and the farmers carry out the PA applications themselves; while in the US Eastern Corn Belt, the fertiliser dealers or co-ops provide these services from the production of maps to the VR applications.

Presentation and Storage of PA Data

The majority of the PA practitioners print out the yield maps and soil maps on paper and only a marginal number of farmers look at the papers on the screen. Yield data are mainly saved on the farmers' personal computers, while soil data are saved either on farmers computers or on co-ops or local advisors' computers, who carry out the soil sampling. An important issue is the backing up of data and only about 49% of the Danish and 53% of the US Corn Belt farmers make back ups of the yield data and 17% of the Danish and 26% of the American respondents make backups for the soil data. The low percentage making backups could be due to lack of awareness regarding data safety among farmers. Further research should examine more specific reasons and potential actions.

Data Ownership and Data Handling

Farmers are very reluctant in entrusting the data storage and data protection to entities outside the farm. Eighty-one percent of the Danish and 78% of the US Corn Belt farmers indicated that they would prefer to store the data themselves, while 88% of the American respondents would prefer not to store the data in a shared Internet-based database. This is an interesting indication on contrary to many potential benefits from data exchange and benchmarking on a regional level. Regardless these figures commercial companies have recently started to provide Internet-based services to farmers, mainly in the USA and Australia. Farmers can send their data to the companies through e-mail and the companies can generate the maps, produce variable rate application maps, as well as agronomic recommendations. The major problem in handling PA data (fig. 2) is the time requirement, where 74% of the Danish and 69% of the American respondents found data handling too time consuming.

Additional surveys have mentioned time requirement as one of the main impediment to PA adoption. Future developments of PA must focus on a better way to handle the large amount of data in order to limit the amount of time for handling the data. More specifically Smith 2002 gave an analogy to PA with biotechnology, where the adoption of biotechnology has been considerably faster than PA. She referred to PA as being time consuming for farmers, while biotechnology is time saving, which is very important in farm management. The American respondents also believed other important factors were lack of technical knowledge and use of software packages. Danish farmers show a similar attitude towards lack of technical and agronomic knowledge, as well as use of software and data transfer (one third of the respondents mentioned all these factors).

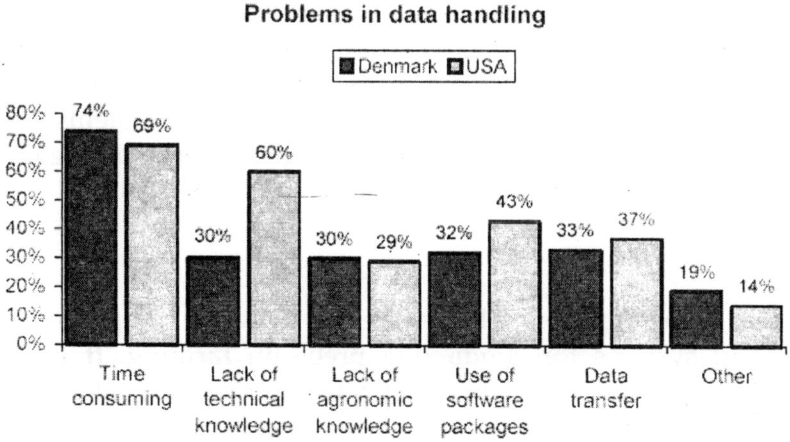

Figure 2: Farmers Concern and Problems about PA Data Handling.

Interesting indications were drawn comparing the responses to the time requirements and lack of technical knowledge by age, years of experience with PA and acreage. The results showed that only years of experience had an influence to responses to time requirements. The American respondents, with 3-5 years of PA experience composed the largest proportion who said that PA was too time consuming; whereas farmers with the least amount of experience found it less time consuming. This is logical, as the first two years of practicing PA is experimental and from the third year farmers start to reasoning about the causes of variability among the accumulated maps. In addition, in the American survey, respondents' age showed to have an influence on responses to the lack of sufficient technical knowledge. Therefore, the youngest category (20-29 years) had the smallest proportion that said insufficient technical knowledge is a problem (17%) and the largest proportion was among the age group (50-59 years) that 63% of the respondents who said insufficient knowledge was a problem. From the personal interviews, one American farmer suggested that "a farmer should have a good structure, good understanding and good record keeping for data tracking, when he first starts collecting data, otherwise, it would be a big mess and the data would not be easily accessible and easy to analyse. The best way of storing the yield and soil data is in a raw format, as it will be easier to be analyzed in the future with new software packages and or overlay them together."

Data Interpretation

An important issue in dealing with PA data is their interpretation. It is not that complicated to collect site-specific data, but the interpretation of these data has been a difficult task for farmers. An interesting comparison has been between the easiness of interpreting yield data and years practicing PA, in the US Corn Belt. About 74% of the respondents who responded that yield maps are "very easy" to be interpreted have used

PA for more than 5 years, while 52% of the 8 respondents who answered "difficult" also belonged to that group. Apparently, even farmers who have been collecting yield maps for a number of years did not find yield maps of practical help. It is possible that the aggregation of maps over successive seasons may cause problems due to temporal variations that many farmers cannot explain or there are still many gaps in agronomic knowledge than issues related to the technology.

Compatibility

A general concern among farmers is hardware and software compatibility and to choose the right technical systems for conducting precision farming. It is in principle the hardware compatibility, which is the most vital obstacle for adoption. Hardware is expensive to change and should in principle be mounted on the tractor for up to 10 years. Several companies have tried to develop their own systems based on ISO-standards but often without chosen solutions that are compatible with other brands. In addition, most of the previous electronic systems for precision farming have not been compatible with a traditional windows-platform. Instead the farm equipment manufactures have developed their own software.

Internet and E-mail Use

Most respondents in Denmark and US Corn Belt mentioned to use extensively the Internet and Email for farm purposes. Only 10% of respondents in both countries have never used the Internet and the majority of them use it daily (about 35%) or weekly (about 32%). High Internet use is consistent with the general observation that farmers who use PA are among the most comfortable with new technologies. Additionally, younger farmers use internet even more on daily basis, as farmers in the age group 20-29, indicated to use the internet for agricultural purposes daily at about 75% in Denmark and 50% in the US Corn Belt. However, the PA practitioners, respondents of the two surveys in Denmark and US Corn Belt, showed that they don't use the internet for PA. The low percentage of those farmers who use the Internet for information about PA may indicate the lack of related web services, and only a small percentage of farmers could recall some PA related internet sites. Concurrent with the use of Internet for information retrieval, widespread use of e-mail was observed (table 1), where about 90% of the respondents in Denmark and the US Corn Belt have used e-mail. More specifically, Table 1. illustrates the e-mail use for specific agricultural purposes.

E-mail is an important means of communication for most respondents. The high rate of e-mail usage can be useful for dissemination of PA information and experience. So far, some companies 9 have established these methods for exchanging information with farmers. In general terms, both Internet and e-mail use show that this group of farmers is among the most technologically advanced in agriculture.

Table 1. E-mail Use for Agricultural Purposes

	% Denmark	% USA
Communicate with other farmers	49	60
Communicate with consultants or co-ops	48	53
Communicate with fertiliser dealers/wholesalers	29	24
Subscribe to Ag websites	37	49
Seek information from commercial companies	24	30
Private correspondence	73	76
Other agricultural purposes	10	29

Information Preferences for Practicing and Investing in PA

The information sources to invest and practice PA are quite different among countries, depending on the agricultural structure and the extension or advisory services. In the UK, farmers acquire information about PA from their private agronomist and from machinery manufacturers. In Denmark, machinery manufacturers and local advisors are the main sources to invest in PA, while local advisors, fertiliser dealers and wholesalers are the main information sources to practice PA. In the US Corn Belt, the most important sources to invest and practice PA were fertiliser companies, information from other farmers, agricultural consultants and the agricultural press. The Internet contribution to information on how to implement PA was low both in Denmark and the US Corn Belt. Figure 2, shows the information sources to practice PA in Denmark and US Corn Belt.

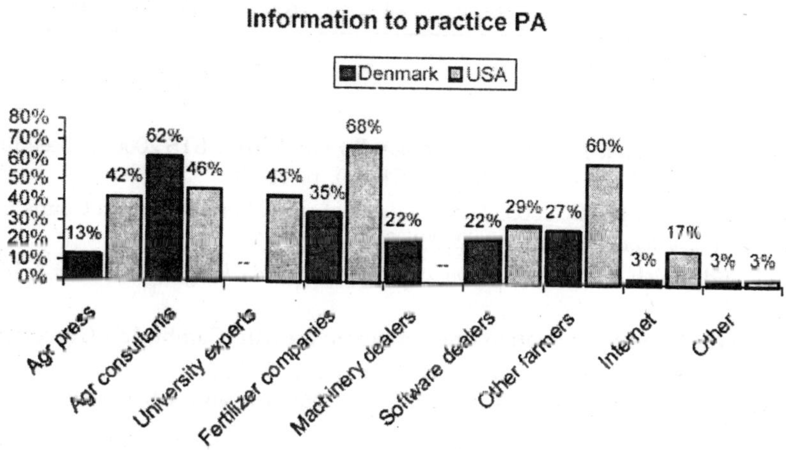

Figure 2. Preferred information sources for practicing PA

Among the Danish respondents, the prominent role of agricultural consultants can be attributed to the extended supportive organisational structure of the advisory service in Denmark. In addition, Danish PA respondents own, to a large extent, their machinery and have a closer relationship with the PA machine dealers. On the other hand, in the

US Eastern Corn Belt, commercial services from fertiliser companies, coops and crop consultants are typical used to implement PA. In 1995, a mail survey conducted by Purdue University with agribusiness specialists and large-scale farmers from the Corn Belt region reported that the most important information sources for PA were agricultural press, university specialists and consultants. That shows a shift in the information sources to practice PA in the USA, from public sources, such as press and university specialists, to private sources, such as fertiliser companies and agricultural consultants.

Table 2 demonstrates the satisfaction from different service providers in PA between Danish ands US Corn Belt farmers. In the American survey, the satisfaction with university specialists and county extension educators was also assessed. About 41% of the US respondents have never used the service, while of those farmers who have used the service, 77% are satisfied with it. For the university specialists, this result may indicate that more attention needs to be devoted to communicating with a broader group of farmers.

Table 2. Satisfaction from different service providers

	Crop advisors		Fertiliser dealers		Machinery dealers		Software vendors		Hardware vendors	
	%DK	%USA	%DK	%USA	%DK	%USA	%DK	%USA	%DK	%USA
Yes	50	52	23	69	39	31	29	33	50	31
No	26	12	16	16	20	37	16	24	29	25
Not used	19	28	56	9	29	26	44	39	11	37
Don't know	6	8	5	6	12	6	11	4	10	7

REFERENCES

Blackmore, S., 2000. "*Developing the principles of Precision Farming*". In ICETS 2000: Proceedings of the ICETS 2000 (China Agricultural University, Beijing, China), p. 11-13.

Blackmore, S., Godwin, R., Fountas, S., 2003. "The analysis of spatial and temporal trends in yield map data over six years". *Biosystems Engineering* (2003), 84 (4), 455-466.

Coombs R. Saviotti P and Walsh V 1987. "*Economics and technological Change*", Rowman & Littlefield, Totowa 1987.

Davies S. 1997. Technical Choise, Innovation and economic growth, Cambridge University Press.

Fountas, S., Ess, D., Sorensen, C. G., Hawkins, S., Pedersen, H. H., Blackmore, S., Lowenberg-Deboer, J., 2005. Farmer experience with Precision Agriculture in Denmark and US Eastern Corn Belt. *Precision Agriculture*. In Press.

Kitchen, N.R., Snyder, C.J., Franzen, D.W., Wiebold, W.J., 2002. "Educational Needs of Precision Agriculture". *Precision Agriculture*, Volume 3, Issue 4.

Lambert, D., Lowenberg-Deboer, J., 2000. "Precision Agriculture Profitability Review". *Site-Specific Management Center*, Purdue University.

U.S. Dept. of Agriculture, 2002. *U.S. Census of Agriculture* (National Agricultural Statistics Service, Washington, DC, USA).

11

Applications of Mobile Mapping System in Agriculture

Mobile Mapping System has strong appliance potential in intelligent transport, precision agriculture, field surveying and environment engineering and so on, so its design and study have been paid more attention in GIS field. Based on the acquirement, compression, storage, transmission, mapping of multi-dimensional agriculture information, precision agriculture is a system engineering integrating information technology and agriculture production. GPS is a real time positioning system and it achieves mobile positioning. Integrating Mobile Mapping System and GPS, the spatial difference information in field can be getting quickly and accurately. Mobile Mapping System supports the display, edit and management of standard vector and grid data, collects and transmits multi-dimension field environment data by Internet and wireless communication equipment, and realises the fast construction and real time update of field information database.

Mobile Mapping System is an information technology that has been developed since late 1980s with the advance and progress in mobile positioning technology, modern communication technology, spatial information technology and embedded technology, etc. Integrating diversiform advanced technologies, Mobile Mapping System is capable of real time data accessing, mapping and spatial data visualisation. Mobile Mapping System not only collects dynamic data about some object in moving, but also manipulates these data in order to make decision efficiently, and make these data be downloaded into a desktop GIS simultaneously by wireless or wire communication as you like. Now Mobile Mapping System has been applied in many fields, such as intelligent transport, precision agriculture, field surveying and environment engineering and so on for outstanding real time supply is very essential in these fields and Mobile Mapping System can do.

Precision agriculture (precision farming, digital agriculture) has evolved from a concept ten years ago into an emerging technology today. Its basic purpose is to manage farmland by the square meter instead of the square mile. It is certain that precision agriculture can be described as the next great evolution in agriculture. With Mobile Mapping System, every inch soil can be monitor and planting data in detail can be acquired anywhere anytime.

MOBILE MAPPING SYSTEM CONFIGURATION

Mobile Mapping System consists generally of three parts, that is moving platform, navigation sensors and mapping sensors. Its configuration can be described as follow (figure 1).

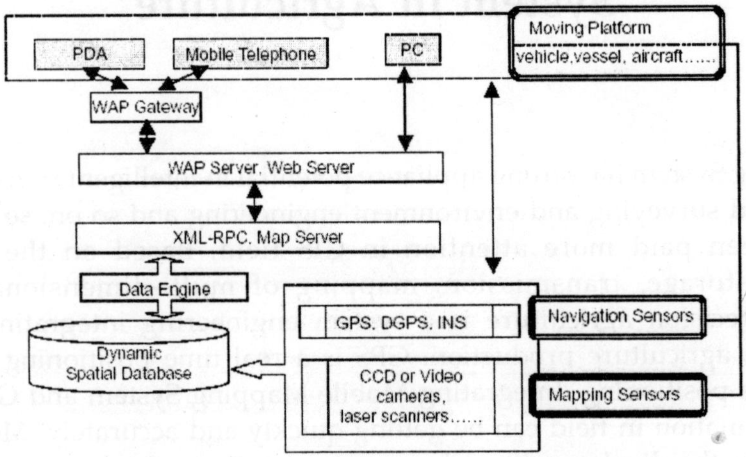

Figure 1: Mobile Mapping System Configuration

The mobile platform is designed to equip data acquisition devices including mapping sensors and positioning apparatus. It may be a land vehicle, a vessel, or an aircraft. Thus the system can observe objects at closer range flexibly. Generally, Mobile Mapping System monitor objects by CCD cameras, video cameras, laser scanners and radar sensors. These mapping sensors are selected according by purpose, but they must obtain high flexibility in data acquisition, more information with less time and effort, and high productivity.

Of course GPS (Global Positioning System) must be connected at the same time in order to record the position and time. Mobile Mapping System is essentially useless without GPS. GPS is developed and initiated by the United States of America. The system has capability of navigation and positioning anywhere on earth, anytime, and under any conditions. Since SA (Selective Availability) was permanently turned off at midnight May 1,2000, the accuracy of GPS data has been increase. And differential GPS techniques can help to obtain high positional accuracies. In addition, vehicle wheel sensors and INS (Inertial Navigation System) are important navigation instruments, too. Integrating with GPS, they can provide both the track of the vehicle and position and orientation information of the mapping sensors. All these data acquisition devices can get plenty useful attribute information. The information will be edited and geo-referenced, then stored in the Spatial Database. Geo-reference process is very necessary because these data will be integrated with other information from multi-sources. Field data collection is always a difficulty for cartographers, surveyors and researchers. The tools available for mapping applications have been bulky in size and weight, expensive, and difficult to learn for a long time. Fortunately, the advances of remote sensing, GPS technology, GIS and some data edit and analysis software drive the field data collection. The advance refers not only precision has been improved, but also the hardware has become smaller, lighter, and cheaper. And the software has become easier to learn, and more inexpensive. So the data collection task becomes easier, more economical and faster to complete.

Mobile terminals should be another important component of Mobile Mapping System. Spatial data can be displayed, edited, or analyzed with them. There are different terminals that can finish such task. They are PDA, Mobile telephones, and Land Phones. These handheld equipments can exchange data with desktop PC via serial or USB ports, and some can do this by Internet access. In many cases, Mobile Mapping System is equipped on vehicle, and above functions can be done on PC. Whatever terminals can interchange data with GPS data, and access data from long-distance database by Internet or wireless communication equipment. So the works need the support of modern communication technologies.

KEY TECHNOLOGIES

Integrates space information, modern communication and intelligent technologies, Mobile Mapping System is designed to collect real time field data with certain positioning information, editing, processing, visualising, and mapping multi-source data. In its working, following key technologies exert important functions.

Intelligent Integration of Multi-Dimension Data

Usually field data derived by Mobile Mapping System at scales have to be combined with ancillary data layers from other sources. The resolution and format of multi-sources should be matching. Moreover it is necessary for GPS data properly matching with the geo-spatial database. Recently the integration method has been much more intelligent for dealing with and integrating quickly and accurately. Such study need further progress though great deal of work has been done on it.

Compression and Decompression Models of Spatial Information

Compression and decompression models study has been discussed for a long time. Almost every spatial database involves the problem because large data files not only take much more capacity of the database, but also at the same time its loading cost much longer time than small files. Now there are many compression and decompression models for spatial information. All of them have their own characteristics and their compressibility is different from each other. Generally higher compression ratio brings more information loss. When selecting the compression and decompression models, the safety, time cost and efficiency are main factors considered.

Efficient Transmission of Spatial Information

Spatial data/information includes not only remote sensing data, survey data, but also graphic data and attribute data, so transmission of the large data is a difficulty especially when telecommuting or data synchronisation between mobile terminals and desktop GIS. Thereby efficient transmission of spatial information is needed imminently. To resolve the problem, bandwidth of the broadband networks should be improved. At present the new networks with 10 G/s transmission speed is building in USA. It is obvious that the transfers of large data depends on broadband networks technology.

Interface of Wireless Communication and Mobile Intelligent Terminal

The capacity and computing ability of mobile intelligent terminal are limited, so it is impossible to store all the data in the limited memory and flash card. Thereby it is absolutely necessarily to transfer the data by Wireless Application Protocol (WAP) from server to user. Accordingly it becomes a key problem to make the interface friendly, and at the same time the data reading speed is another issue settled imminently. Thereby the study on interface of wireless communication and mobile terminals has been paid on much more attention.

APPLICATION IN PRECISION AGRICULTURE

Precision Agriculture

Precision agriculture is a new agriculture technology system development quickly in recent years. Integrating agronomy, geography, biology, agrology, botany, geo-spatial science, precision agriculture can be defined as a comprehensive system designed to help farming. It involves various problems in crop planning and includes tillage, planting, chemical applications, harvesting, and post harvest processing of the crop. Precision farming is a pro-active approach that reduces some of the risk and variables common to agriculture, so we think that it has the potential of optimising cost and ecological effects through the application of crop information, advanced technology and management practices. Accordingly, how to get these information and data quickly and accurately becomes the foundation of precision agriculture construction, while how to manage the information and make decision quickly and intelligently is the key technology in precision agriculture. It is obvious that information technologies are foundation of precision agriculture, and positioning, timing, mapping and analysis are most important among them. Accordingly Mobile Mapping System can help agriculturist with a new capability of gathering information for implementing decision-based Precision Agriculture.

Mobile Mapping System in Precision Agriculture

For precision farming, information technologies are so essential that Mobile Mapping System is important or beneficial. Equipped with mapping sensors and navigation sensors, Mobile Mapping System can collect field data anytime, anywhere, in any manner. At the same time, Mobile Mapping System can connected with a GIS or combine GIS software in its mobile terminals directly. Thus Mobile Mapping System can monitor planting process all along. Moreover it can contrast multi-temporal data collected or stored in database and find what changes occurred, where, when and how, then give an efficient plant plan. What function the Mobile Mapping System has in precision agriculture is narrated in Figure 2.

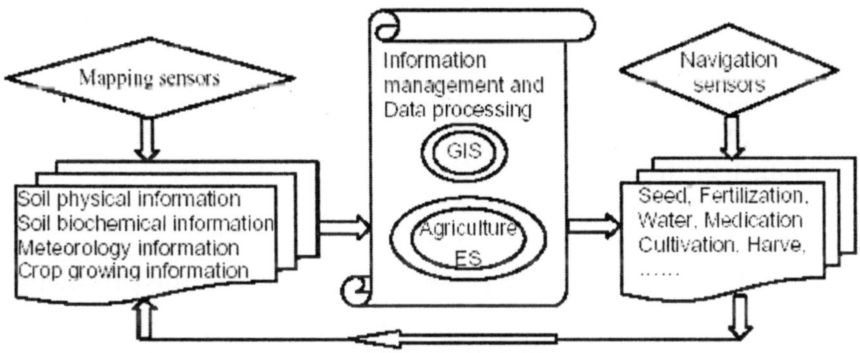

Figure 2: Function of Mobile Mapping System in precision agriculture.

Improve the Accuracy of Soil Sampling

Soil conditions and soil quality influence crop growth greatly. Usually how often do farmers fertilise the soil or water the land depends on the soil conditions and plant type. So it is a basic work to master the soil conditions. From the view of precision agriculture, soil is different meter by meter. Thereby soil sampling becomes the foundation step. The accuracy of soil sampling is requested higher. Accuracy of soil sampling refers mainly to accurate degree of the position information where the soil samples were taken. With the accurate knowledge about the coordinate location of the soil samples, a soil data layer can be developed accordingly. With the accurate position information, navigating back to those locations for re-sampling is possible. Mobile Mapping System has equipped navigation sensors, such as GPS and INS, so the time and position information can be recorded at the same time when soil samples are taken.

Plant Growth, Diseases and Insect Pests Monitor

Generally precision agriculture is constructed in a large area. Hereby it is impossible for all plants growing in like manner. Sometime only a single part suffers the diseases and insect pests. Then it is unnecessary and even harmful for all plants to spray pesticide or other medicine in the same way. Different solution should be taken according to the real conditions of plants. Mobile Mapping System can finish the monitor task of all the plants. Using the scanners, the peculiar plants can be found quickly, and its position also can be recorded. In fact satellite remote sensing data also can provide such distinction information. Compare with Mobile Mapping System, its spatial range is much larger but space resolution is lower and constrained by satellite calendar.

Analysis of the Crop and Field Information

The purpose to get so much data about the soil, plant and so on is to master the planting conditions and to make decisions for all planting process. Accordingly GIS and Agriculture Expert System are usually imported in precision agriculture in order to edit, process, integrate and analyze the crop and field information and get corresponding resolve scheme. Mobile Mapping System may Load multi-source information and different types of agricultural data to master the conditions of the field. Comparing the information or data, relationships within and between data sets can be found. That is the relationships among different factors can be made certain. By the relation, agronomist or Agriculture Expert System can make production plan. After the production plan has been made, farmers can command the farming machines to work automatically, for these machines has been equipped intelligent implement and positioning devices. Moreover the entire process of one year farming also can be recorded and evaluated, and the experience can be analyzed to help next year work.

REFERENCES

Michael Rasher, "The use of GPS and mobile mapping for decision-based precision agriculture", http://www.gisdevelopment.net/procpdf/133-141.pdf

Naser El Sheimy, Kaichang Di, Ron Li, "Report on the 3rd International Symposium on Mobile Mapping Technology", Cairo, Egypt, 2001

Khurram Niaz Shaikh etc. "GPS-Aided-INS for mobile mapping in precision agriculture", http://www.gisdevelopment.net/technology/gps/ma03194.htm

Ashok Wadhwani. "Recent advances in mobile GPS/GIS mapping technology", http://www.gisdevelopment.net/technology/mobilemapping/techmp005a.htm

12

Agroinformatics

Agricultural data is currently generated by multiple agencies in multifarious ways; both conventional field surveys based as well as advance information technology based. Some of the prominent agriculture data publishing programmes in India are: CAPE (Crop Acreage and Production Estimation), FASAL (Forecasting Agricultural output using Space, Agrometeorological and Land based observations). Federal Agricultural department generates the data by field sampling surveys. Industrial houses send their own field team to assess the acreage and production data. Agencies like Agriwatch, CSE are also gathering Agricultural Intelligence (AI) data from multiple sources. Aiming to resolve such issues of vagaries in the AI data, RMSI has established a geospatial approach to rationalise a competent methodology for SCM (Supply Chain Management) which can benefit farmers, traders, exporters, industrial, government, and federal agencies to combat the exports. This chapter highlights this through a specific case study conducted by RMSI for estimating crop acreage estimation, crop yield estimation and production estimation for various rice exporters.

UTILITY OF BASMATI AGRICULTURAL INTELLIGENCE (AI) DATA

RMSI understands that AI data generated is used in different ways:

1. *Industries:* Use this data mainly for procurement and supply chain management
2. *Boards:* Use this data to streamline supply chain as well as to fix the price in the market
3. *Insurance companies:* Create an insurance product out of the agricultural yield data

However, the conventional data does not suffice for many of the users. They need agricultural data modeled in such as changes in the cropping pattern from the last year, comparative analysis of the last year data vs. the current year, production pertaining to respective mandis, settlement packages for farmers without affecting the profit margin of the insurance companies, etc. To create such intelligent data, RMSI followed two different aspects, namely:

1. Geospatial data validity of comparing Mandi data with remote sensing based outputs in rice production
2. Supply chain management methodology evolved from the above survey

Study Area

The study area covers major rice growing districts from the Indian Ganges flood plains. This includes 13 districts each in Punjab and Haryana, 29 districts in Uttar Pradesh, 4 districts in Uttarakhand and 2 districts of Jammu & Kashmir. Geographically, this spreads across 25°83′ North to 33°07′ North latitude and 73°87′ East to 81°86′ East longitude covering an area of about 189,000 sq km.

Input Data

The study entailed collection, procurement and analysis of primary and secondary sources of information. The broad classification of agricultural acreage over the entire region was carried out using IRS P6 AWiFS satellite images with spatial resolution of 56m. Information, from regional to local, were extracted using medium and high resolution satellite images of IRS series 1C, 1D, P6 LISS III and LISS IV with spatial resolution of 23.5m and 5.6m, respectively. Secondary sources of information like Survey of India Toposheet of India Toposheet on 1:50,000 scale, district maps for the study area were used as reference maps. RMSI also collected primary information for ground truth, field validation, sample based farmer survey and Mandi data through field based-surveys in all the districts. Secondary information was also collected from district agriculture and state agriculture boards/offices for reference.

METHODOLOGY

Major components of compiling a methodology for SCM included Market survey and assessment and Agriculture and Land Resource Mapping. While market survey is done through direct fi eld authenticated data, agriculture and land use map was envisaged through interpretation of remote sensing satellite images. With a cursory look at mandi information, it is apparent that both quantity and expected time of arrival information are vital. RMSI selected a sample district mandi in each of the Basmati growing states. Mandi arrival data (quantity) and time of arrival was collected from important mandis and market board at block/tehsil level and mandi board at district and state level. It was observed that Rice and Bajra (Pearl Millets) crops come early in the market i.e., September–October and Basmati and Sugarcane varieties come later in November–December months. To envelop this variation in time, a thorough, periodic and regular survey was carried out.

Geo-spatial Data Validity

In conventional ways, after the estimation, second level sample surveys are carried out (crop cutting experiments) or remote sensing based ground verification is done. However, the data produced, often does not sync with the final output data produced from mandis or markets or final government figure. This, in turn, leads to the question of authenticity of data produced from remote sensing. RMSI undertook a hybrid approach to compare and assess the accuracy of remote sensing-based production estimation of Basmati rice for kharif 2005 against the mandis arrivals and other related sources at the end of the Kharif season-2005. Mandi arrival data of rice and Basmati was collected for Haryana and Punjab, whereas in Uttar Pradesh and Uttaranchal only total rice data was available and the same was collected. The survey experts also collected data from rice mills, state marketing boards of respective states and the Agents (Arthias). Analysis was carried out based on information gathered from various sources like discussions with respective mandi officials, mandi agents (Arthias), and rice millers to arrive at the conclusion. Based on comparative analysis between Kharif 2005 estimates derived using remote sensing approach and Mandi arrivals (as well as allied sources) as on 31st March 2006 and considering certain calculated and logical assumptions and limitations, it is concluded that the remote sensing based estimated results for Kharif 2005 is matching up to an accuracy of 90% to 94% in the states of the study area. As a by-product, an interesting result on RS data being always higher than total supply chain sources confirms the reliability of this data.

ADVANCES IN REMOTE SENSING FOR AGRICULTURE

Remote sensing technology has seen many changes in the past five years. The major changes are that from satellite altitudes we are or will be able to (1) image or see with more detail, a smaller piece of land, (2) define more precisely the specific colors or light responses reflecting off of the field and (3) obtain data on a regular interval of every other day or every 5-7 days. These changes make for real advances to agriculture as we need to be able to view those small areas in the field that are giving us problems, determine what the problem is within a field by interpreting remotely sensed data and receive data/ information on a regular interval.

SPATIAL RESOLUTION: THE SPACE BETWEEN MY DATA POINTS

When you view your yield image map, you are looking at about 750 to 1,500 data points per acre depending on how fast you were traveling and how often (1, 2 or 3 seconds) you made a yield measurement. The area that one can see on an image, whether yield monitor or remotely imaged is called "spatial resolution". Spatial resolution in satellite data collection is improving. With current satellites, one can see areas that are 30 meters ´ 30 meters (4.5 measurements/acre), 20 ´ 20 meters (10 measurements/acre) and 10 ´ 10 meters (40 measurements/acre). With future satellites, we will be receiving data that have a variety of spatial resolutions or pixels (short for picture elements) that in some cases will be as detailed as 1 ´ 1 meter or over 4046 data points per acre.

Table 1: Conversion of Spatial Resolution from Images to Data Points on the Ground.

Resolution (meters)	Data Points (Pixels)/Acre	Points/Hectare
1000	0.004	0.01
80	0.6	1.56
30	4.5	11.1
23.5	7.3	18.1
20	10	25.0
15	18	44.4
10	40	100
5	162	400
4	253	625
3	450	1111
2	1012	2500
1	4046	10000

In terms of sensor technology, we are seeing improvements in spatial resolutions that allow one to see great detail (Table 2). At recent conferences, we have learned that there is a potential for over 50 land observing satellites to be launched before the year 2007. These satellites which are both government and commercial will have a large range of spatial resolutions from 1 meter to 1 kilometer. The highest resolution is proposed by Space Imaging Inc. and Earthwatch (QuickBird satellite) who plan to launch satellites in the next few years that will have 1 meter panchromatic and 4 meter digital data. EarthWatch (EarlyBird satellite) and other companies like Resource 21 and GER are promising data in the 5 to 15 meter spatial range.

Table 2. Remotely Sensed Data Choices of the Future and Some Characteristics.

Satellite Platforms

Spatial Resolutions:

1, 2, 3, 4, 5, 10, 15, 20, 23.5, 30 meters

Spectral Resolutions:

5, 10, 50, 100 nanometers

Temporal Resolutions:

1-5 days, 8-10 days, 14-16 days

Aircraft Platforms

Sensors similar to satellite platforms

Greater flexibility in obtaining coverage

Fly under the clouds!

Tractor Platforms

Potential sensors for organic matter content, pH, soil texture, soil moisture

Sensor measurements relating to specific tasks

Work under a variety of conditions

Several companies are providing sensor coverage by airplane from a variety of altitudes at the time that the grower would like to see what is happening to the crop. The other choice is that companies are developing sensors that can provide specific measurements for specific elements or conditions. Purdue University has patented an organic matter sensor that can be used on an implement such as a corn planter or on a lime/fertilizer truck. Other sensors are being developed to measure soil texture, pH and soil moisture which show promise of being a high growth business opportunity.

SPECTRAL RESOLUTION: THE "VARIOUS COLORS" OF MY DATA POINTS

When remote sensors talk about the variation of light energy and its measurement, they call it spectral resolution. Improvements in color differentiation and refinement of measurement response have been in measuring a smaller portion of light energy or spectral wavelength bands measured in nanometers. The sensors placed in future satellites will take advantage of technology that will allow for measurements of narrower bands from 100 down to 5 nanometers and therefore a better measurement of the different colors and of areas that the human eye cannot see in the near infrared. Currently satellites like Landsat have 7 wavelength bands and SPOT has 4 wavelength bands with bandwidths of about 50 nanometers. TRW will launch the Lewis satellite during this next year under a NASA contract with 384 wavelengths. The 5 nanometer bandwidth off the Lewis sensors will certainly provide a more detailed look at the reflective nature of light coming from soil and plant conditions. Other companies will learn from this data in their selection of specific wavelength bands for specific measurements of plant conditions such as stress, nutrient deficiencies and similar variations effecting crop yield.

TEMPORAL RESOLUTION: THE NUMBER OF TIMES THAT I CAN SEE MY DATA POINTS

The same field or location on repetitive dates, using temporal resolution. The variation that occurs over time or "change detection." With current satellite data, one has the potential of receiving data every 2-6 days (SPOT) or every 16 days (Landsat). Future opportunities will be either 2, 3 or 5 days depending on the company. The value of receiving repeat data would be in identifying a change where one could perform an activity that would correct or improve the change. If you are irrigating, it may be in changing the rate or amount of water, or adding nutrients with the water. If the change is due to some stress factor such as weeds, insects, diseases, or similar pests, one would need to decide if correcting the problem through application of a pesticide would be economical. Growers are also interested in estimating the yield of their crop prior to harvest. Remotely sensed data, especially with a temporal view, can give valuable clues to the potential yield; however this will vary by growth and maturity of the crop. Not all crops will provide high correlation of vegetation mass to yield. Variation in corn varieties, for example, has shown that some varieties have more vegetative cover than others and may not be related to actual yield. Much work needs to be done in this area.

REMOTE SENSING, GEOGRAPHICAL INFORMATION SYSTEM AND PRECISION FARMING IN INDIA: OPPORTUNITIES AND CHALLENGES

Remote Sensing in Agriculture: Present Scenario

Remote sensing techniques play an important role in crop identification, acreage and production estimation, disease and stress detection, soil and water resources characterisation and also by providing required inputs for the following: generation of land and water resources developmental plans, bringing 479 additional land into cultivation through mapping and reclaiming wastelands, increasing the irrigation potential through ground-water prospects mapping; crop-yield and crop-weather models, integrated pest management, command area management, watershed management, agrometeorological services, precision farming, etc. Remote sensing applications to agriculture have grown to a stage where such inputs are being used for number of policy level decisions related to food security, poverty alleviation and sustainable development in the country. Decision on buffer stock of food grains could be based on pre-harvest crop acreage and production estimation while the ground water prospects maps serve as the major source of information in ensuring drinking water and other needs in rainfed and less favoured areas. Nation wide wasteland, land use, land cover and soil mapping has helped in expanding and intensification of agricultural activities and also in identifying the land capability classes and crop suitability indices.

CROP ACREAGE AND PRODUCTION ESTIMATION

In India the use of space-borne remote sensing data for crop acreage estimation and production forecasting was experimented in early 1980s in selected districts for wheat, rice and groundnut. The promising and encouraging results of this initial study led to an attempt to estimate state-level wheat acreage using Landsat MSS data for Haryana and Punjab in 1985-86. The results were encouraging and the project, namely, 'Crop Acreage and Production Estimation' (CAPE) was launched covering wheat, rice, groundnut and *rabi* sorghum in selected major growing states/districts. Since availability of optical data is a concern in monsoon seasons, the use of data from active sensors, such as SAR of RADARSAT was operationally used for *kharif* rice in 12 districts of Karnataka. The microwave data, which has all weather capabilities, has shown that the rice crop can be discriminated at better than 90% accuracy, which can aid in early detection with multiple forecasting. Apart from using the single-date high-resolution satellite data to provide crop acreage estimates at district-level under CAPE, the multidate WiFS data (coarse resolution and high receptivity) is used to explore the possibility of making national-level forecasts. The procedure uses a national-level sampling frame and coarse sample segment grids to make multiple forecasts as well as inter-seasonal crop growth differences using multi-year WiFS data. A comprehensive software package called 'CAPEWORKS/CAPEMAN' has been developed enabling end-to-end analysis culminating in generation of production statistics. The remote sensing techniques have also been deployed in assessing the horticultural crops. For example, acreage estimation of mango and banana plantations in Krishna district of AP and Thiruchirapalli district in Tamil Nadu, respectively has also been successfully carried out with an accuracy of 94% in a joint venture with the Indian Institute of Horticultural Research. Mapping of sugarcane has been carried out to aid various users and cooperative sugar mills in accurately estimating the acreage. This is also done at cadastral level enabling the government to identify those beneficiaries (farmers) who are not paying the tax. Efforts made on estimating the onion acreage in Nasik district, Maharashtra yielded results to an accuracy of 75% because of smaller field size, peculiar leaf structure, continuous cultivation and spectral mixing with other crops.

From the experience gained on implementation of CAPE project and also to meet the requirements of timeliness, accuracy and coverage of crops, an integrated concept of Forecasting Agricultural output using Space, Agrometeorology and Land based observation (FASAL) has been evolved. A National Crop Forecasting Centre is being established by the Department of Agriculture and Cooperation, Govt. of India to execute the project. The concept of FASAL, thus strengthens the current capabilities of early crop season estimation from econometric and weather-based techniques with remote sensing mid-season assessments that can be supplemented with multi-temporal coarse resolution data based analysis. In the later half of crop growth period, direct contribution of remote sensing in the form of acreage estimates and yield forecasts is available. However, in this case also, the addition of more extensive field information and weather inputs would increase the forecast accuracy. India is also planning for some exclusive satellites to provide data specific to agriculture. Some of the current and future missions of Indian space programme related to agriculture are presented in Table 1.

Table 1: Major Indian remote sensing missions for agriculture.

Mission	Year of launch	Sensors
IRS-IA, IB	1988	LISS-1 (72.5 m resolution;148 km swath)
	1991	LISS-II (36.25m resolution;142 km swath)
IRS-P2	1994	LISS-II (36 m resolution; 142 km swath)
IRS-IC, ID	1995	PAN(5.8m resolution;70 m swath)
	1997	LISS-III (23.5m, 70.5 m resolution; 141 km, 148km swath) WiFS(188.3 m resolution; 774 km swath)
IRS-P3	1996	WiFS (188.3 m resolution; 774 km swath)
TES	2001	PAN (1m resolution;14 km swath)
RESOURCESAT-I	2001	LISS-IV (6 m resolution: 25 km swath / LISS-III /23 m resolution; 140 km swath/ AWiFS /80 m resolution;800 km swath)
CARTOSAT-1	2002	PAN Stereo (2.5 m resolution;30 km swath)
CARTOSAT-2	2002/03	PAN Stereo (1 m resolution;12 km swath)

Apart from the above, India is also planning a series of INSAT and METSAT satellites to improve the applications in meteorology, weather, disaster management and atmospheric studies which have indirect applications to agriculture. The METSAT series of satellites will be carrying some additional sensors which will be useful for agriculture and forest applications apart from the VHRR (2 km resolution in visible and 8 km resolution in thermal infrared and water vapour bands) for meteorology and weather applications. The application of remote sensed data in the study of crops, soils, water, etc. presented in Table 2:

Table 2. Important remote sensing application projects in Agriculture.

S.No	Project	User organization
1.	Crop Acreage and Production Estimation (CAPE)	State Departments of Agriculture/Ministry of Agriculture
2.	Cotton Acreage and Condition Assessment (CACA)	Ministry of Textiles.
3.	Forecasting of Agricultural output based on Satellite, Agrometerorology and Land based observation (FASAL)	State Department of Agriculture
4.	Nation wide mapping of soil resources on 1:250000 scale	Ministry of Agriculture
5.	Mapping of saline/alkanine soils of India	Central and State Agricultural Departments.
6.	National wasteland mapping on 1:1 Million scale/1:50,000 scale	Wasteland Development Department /District Authorities
7.	Nationwide mapping of landuse/land cover for agroclimatic zones on 1:2,50,000 scale	Planning commission
8.	National Agricultural Drought Assessment and Management Systems (NADAMS)	Department of Agriculture and Co-operation, State Agricultural Departments and District Authorities.
9.	Forest cover mapping for the entire country	Department of Forest, Forest Survey of India, State Forest Departments.
10.	National drinking water technology mission on 1:2,50,000 scale	Ministry of Rural Development, Ground Water Department
11.	Satellite monitoring of command areas for crops, salinity/alkalinity and waterlogged areas	Central Water Commission / State Command Area Development Authorities.
12.	Integrated Mission for Sustainable Development (IMSD)	Planning Commission, Ministries of Rural Development, Agriculture, Environment and Forests, State governments.

GEOGRAPHICAL INFORMATION SYSTEM (GIS): IN AGRICULTURE

Integration of the land and water resources and identification of the constraints/ ecological problems at the micro level will help in identifying the location specific solutions through the effective use of remote sensing based resource information combined with other socio-economic data using GIS. This is being done by survey of resources at different scales using traditional and remote sensing techniques, collection of collateral information like slope, topography etc, preparation of a set of resource maps (hydrogeomorphology, soils, landuse/cover, surface water/drainage/watershed etc.) and generation of action plan maps giving site specific recommendations for development and management of agriculture, ground water recharge, fuel and fodder as well as for soil conservation/reclamation and afforestation. In one of the experiments at NRSA, sustainable action plan for upper Machkund Watershed falling in tribal areas of Visakhapatnam district, Andhra Pradesh has been generated through GIS techniques. GRAM ++ is an indigenously developed Geographical Information System (GIS) software package for storage, analysis and retrieval of geographic information relevant to the task of local level planning (GRAM++—Geo-referenced Area Management Package A Tool For Local Level Planning, Natural Resources Data Management System, Department of Science & Technology, Government of India). The package has been put to use in a variety of applications like watershed management, waste land reclamation, land capability analysis, soil erosion assessment, energy budgeting, location/ allocation of facilities, and hazard zonation studies. These applications have been conducted and functionalities tested by Karnataka State Council for Science and Technology, Bangalore; Technology Informatics Design Endeavour, Bangalore, National Atlas and Thematic Mapping Organization, Calcutta, National Water Development Agency, New Delhi and several other universities and R&D institutions. Under NNRMS, Dept. of Space, a Natural Resources Information System (NRIS) is being developed for policy makers to ensure the optimum utilization of natural resource information. This information system will enable to update existing natural resource information and integrate with socioeconomic data. GIS forms the core of NRIS for data storage, retrieval, data integration and analysis. It will also be linked with other existing information systems at district/state level to provide an efficient and powerful tool for management of resources. NNRMS has prepared a document on the node design and standards to be adopted under NRIS for all themes like soil, land use, water, geomorphology, socioeconomic data, infrastructure facilities, etc.

Thus, the potential of the GIS is vast to improve the resource management for enhancing the productivity on a sustainable basis in an eco-friendly manner. This involves collection and formatting large database with co-operation from line departments/institutions for deriving the maximum benefits out of this technology.

Precision Farming

Précising farming is an emerging concept in modern agriculture. It is a micro management system to arrive at improved agricultural and land management decisions that result from using information delivered by geospatial technologies. In other words it is "Digital Agriculture" involving very large scale farm level mapping, comprehensive data base creation on required resources generated through space based inputs and field observations and making a detailed plan of work for maximizing the yield and reducing the cost on inputs using the decision support system. The precision farming data base generally includes:

1. Crop characteristics like, stage of the crop, crop health, nutrient requirement, etc;
2. Detailed soil layer with physical and chemical properties, depth, texture, nutrient status, salinity and toxicity, soil temperature, productivity potential, etc.,
3. Microclimate data (seasonal and daily) about the canopy temperature, wind direction and speed, humidity etc.,
4. Surface and sub surface drainage conditions;
5. Irrigation facilities, water availability, and other planning inputs of interest.

With the advent of differential Global Positioning System (GPS) with high accuracy, there are possibilities of automating the farm operations like tillage, planting, fertilizer applications, pesticide/herbicide spraying, irrigation, harvesting and other mechanized cultural operations. High resolution Digital Elevation Models (DEM) may also form a component of the database, which provides the appropriate description of the topography to support soil moisture and fertility models for decision making.

Precision farming enables micro-management concepts that include the ability to appropriately manage every field operation at each location in the field, if it is technically and economically advantageous to manage at that level. The concept of precision farming is picking up very fast in developed countries due to large farm holdings and fully mechanized agricultural operations. With the present day technological developments and further availability of higher resolution multi spectral sensor data, there is scope for adopting precision farming for cultivation of high value/Commercial/fruits/flowers/ vegetables etc. in developing countries including India. The application of remote sensing and GIS techniques in the management of agricultural resource are increasing rapidly due to improvement in space borne remote sensing satellites in terms of spatial, spectral, temporal and radiometric resolutions. Many of conventional approaches of handling multithematic information to arrive at optimal solutions are being computerized using GIS utilities. Keeping in view the development in satellite, computer and communications technologies, the following opportunities and challenges exist in the application of remote sensing GIS and precision farming technologies in India.

OPPORTUNITIES

1. Prioritization of macro/micro watersheds for implementation and impact assessment of watershed projects at national, state, district, taluka and hobli levels.
2. Forecasting of outbreak of pests and diseases based on soil water status and plant stress indicators in crops such as paddy, wheat, sugarcane, cotton, chilli, & pigeonpea, etc.

3. Development of decision support system for precise management of resources at farm level at least in commercial/fruit/flower crops to begin with.
4. Airborne SAR data utilization for identification of *kharif* crops and development of procedures for canopy backscatter models for identification and yield prediction.
5. Soil mapping at cadastral scale using high resolution spatial, spectral and radiometric resolutions.
6. Quantification of soil loss.
7. Detection of water logging due to rising ground water table.
8. Delineation of salt-affected soils in black soil and sandy regions.
9. Soil moisture estimation and mapping using micro wave/optical/thermal remote sensing techniques in surface and root zone depth.
10. Land surface temperature estimation using thermal and microwave remote sensing techniques.
11. Hyper spectral studies on soils to establish quantitative relationship between spectral reflectance and soil properties.
12. Development of digital techniques for a variety of applications using GIS techniques. For e.g., soil suitability to crops, land capability classification and land irrigability assessment etc.
13. Preparatory activities towards hyper spectral data utilization for understanding the plant processes and development of spectral response models for stress detection.
14. Improved yield models by integration of biophysical simulation and regional level crop models.

CHALLENGES

The application of Remote Sensing, GIS and precision farming techniques in the management of agricultural resources are increasing rapidly due to improvements in space science supported by computer and communication technologies. The following challenges need to be addressed in the application of these technologies.

1. Identification of crops and estimation of area and production of short duration crops grown in fragmented land holdings, in particular during *kharif* season.
2. Forecasting of droughts/floods.
3. Detection of crop stress due to nutrients, pests and diseases and quantification of their effects on crop yield.
4. Automation of land evaluation procedures for a variety of applications using GIS techniques.

5. Information on sub-surface soil horizons.
6. Extending precision farming database to smaller farm size and / or diverse crops/ cropping systems.
7. Developing decision support systems for management of biotic and abiotic stresses at the farm level.
8. More accurate yield models
9. Estimation of depth of water in reservoirs and quality assessment of ground water.
10. Better than 1m contours for watershed development plan at the micro level.
11. Use of remote sensing and precision farming technologies in intercropping/multiple cropping situations.
12. Identifying ways and means of reducing the cost of RS, GIS and precision farming technologies and time gap in collection, interpretation and dissemination of data to enable their usage on a large scale. A successful example in this direction is that of hand-held radiometer developed by Optomech Engineers Hyderabad in collaboration with space application center, ISRO Ahmedabad for standardizing the spectral signature insitu for inter pressing the RS Data.
13. Convincing evidence to prove the utility and economic viability of these technologies so as to mobilize support for R & D work.
14. Human resource development to hasten the process of large scale use of unexplained and cutting edge technologies that have tremendous scope and potential.

REFERENCES

Agro-Informatics 2001. "Proceedings of the first National Conference on Agro-Informatics (NCAI)" (V.C.Patil, G.S.Dasog, A.R.Alagawadi, K.C.Shashidhara and D.P.Biradar), INSAIT. Dharwad

Gowrisankar, D. and Adiga, S., 2001, *"Remote sensing in agricultural applications: An overview"*. PP: 9-14, In: Proc, The First National Conference on Agro-Informatics (NCAI), INSAIT. Dharwad.

Venkataratnam, L., 2001, *"Remote sensing and GIS in Agricultural Resources Management"*. pp: 20-29. In: Proc, The First National Conference on Agro-Informatics (NCAI), INSAIT. Dharwad.

13

Use of Remote Sensing in Rice-based Farming

Increased rice area, however, has led to the development of high water tables and risk of large tracts of land becoming salt-affected in southern NSW. These concerns have led to various environmental regulations on rice in the region, culminating in 1994 when restrictions on rice area, soil suitability, and water consumption were fully enacted. Strict environmental restrictions in combination with large areas of land make the management of this region a difficult task. Land managers require, among other things, a way of regulating water use, assessing or predicting crop area and productivity, and making management decisions in support of environmentally and economically sustainable agriculture. In the search for more time and cost effective methods for attaining these goals, while monitoring complex management situations, many have turned to remote sensing and Geographic Information System (GIS) technologies for assistance. The spectral information and spatial density of remote sensing data lends itself well to the measurement of large areas. Since the launch of LANDSAT-1 in 1972, this technology has been used extensively in agricultural systems for crop identification and area estimation, crop yield estimation and prediction, and crop damage assessment. The incorporation of remote sensing and GIS can also help integrate management practices and develop effective management plans. However, in order to take advantage of these tools, users must have an understanding of both what remote sensing is and what sensors are now available, and how the technology is being used in applied agricultural research.

CONCEPT OF REMOTE SENSING

Remote sensing is the acquisition of digital data in the reflective, thermal or microwave portions of the electromagnetic spectrum (EMS). Measurements of the EMS are made either from satellite, aircraft or ground-based systems, but it is characteristically at a distance (or "remote") from the target. Due to the large spatial extent of the areas considered for the CRC for Sustainable Rice Production Project 1.1.05, this report will focus on data gathered from satellite remote sensing systems. Remotely sensed images are recorded digitally by sensors on board the satellites. An example of a satellite operation is shown in Figure 1. The satellites vary in height above the Earth's surface from approximately 700 km, which orbit the earth, to some 36 000 km, which are geostationary above the equator. The images can be manipulated by computers to highlight features of soils, vegetation and clouds. Each pixel, or picture element, contributing to the image is a measurement of a particular wavelength of electromagnetic radiation at a particular spatial scale for a particular location at a specific time. The most common display of remotely sensed data is a single overpass, which non-remote sensing specialists may think of as a 'satellite photo'.

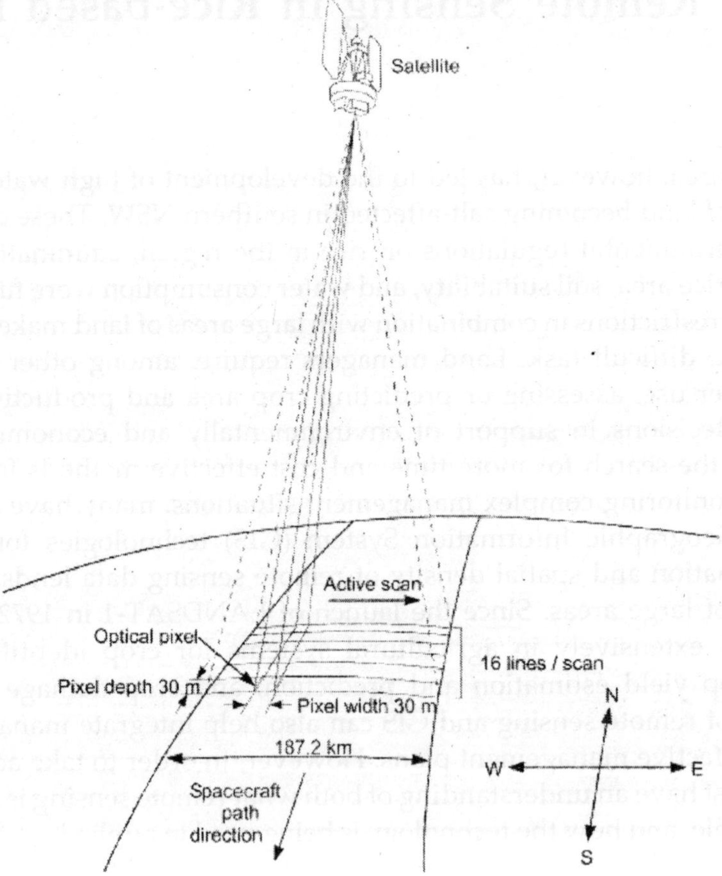

Figure 1. Schematic of satellite operation.

When dealing with remotely sensed images, the extent, resolution, and density of the spectral, spatial and temporal characteristics need to be considered. Spectral extent describes what portion of the EMS is being sampled (e.g., is it just visible or does the range extend into the thermal). Spectral resolution refers to the bandwidths in which the sensor gathers information. Spectral density indicates the number of bands in a particular portion of the EMS (e.g., hyperspectral sensors have higher spectral density than broadband instruments). Spatial extent is the area covered by the image, while spatial resolution refers to the smallest pixel or picture element acquired. Spatial density refers to the amount of area measured by the sensor, so the spatial density for remotely sensed data is complete, while the spatial density of, for example, rainfall stations which are sampled at explicit points, would be incomplete. This means, for the extent of the image, remotely sensed data are a 'census' at a particular spatial scale recorded at a specific time. Temporal extent is the recording period over which the data is available. For some systems, data has been recorded for over 25 years. Temporal resolution is the time that the data is acquired over. Remote sensors usually have a low temporal resolution (i.e., a matter of seconds), while rainfall data recorded at meteorological stations usually have a very high temporal resolution (i.e., they record almost continuously). Temporal density is the repeat characteristics of the satellite and, for some applications, greatly influences the availability of cloud free data.

Remote sensing of the land surface occurs at wavelengths of the EMS where the light can pass through the Earth's atmosphere with no, or little interaction with the atmosphere. These bands of the EMS are called 'atmospheric windows' and refer to the spectral extent in which radiation reaching remote sensing instruments carries information about the Earth's surface conditions. These 'atmospheric windows' are defined by the transmittance of the constituents of the Earth's atmosphere. There are some gases, which absorb all electromagnetic radiation in certain wavelengths preventing these areas of the EMS being used for remote sensing. Figure 2 shows the divisions of the EMS.

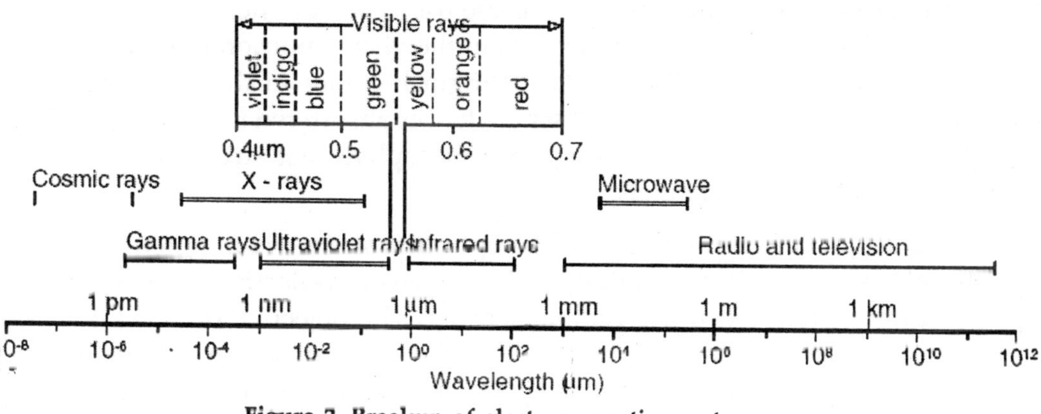

Figure 2. Breakup of electromagnetic spectrum.

One basis of remote sensing is that different land covers have different spectral properties. Figure 3 shows idealised reflectance plots for two vegetation, soils and water types, respectively. Different surfaces also have varying responses in the thermal and microwave atmospheric windows of the EMS. Image processing approaches utilise these differences to extract information relevant for rice-based irrigation systems.

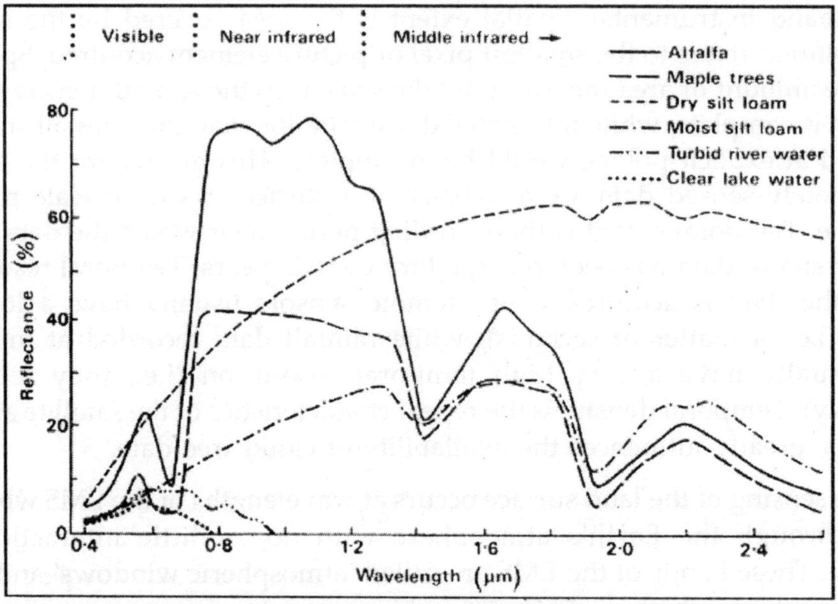

Figure 3. Idealised reflectance plots for different land cover types.

Reflective Remote Sensing

The reflective portion of the EMS ranges nominally from 0.4 to 3.75 micro meters (μm). Light of shorter wavelength than this is termed ultraviolet. The reflective portion of the EMS can be further subdivided into the visible 0.4 to 0.7 μm, near infrared (NIR) 0.7 to 1.1 μm, and mid infrared 1.1 to 3.75 μm. It is in the visible portion of the EMS. Different surface reflective properties allow us to distinguish colour in the visible region of the EMS. Chlorophyll pigments that are present in leaves absorb red light. In the NIR portion, radiation is scattered by the internal spongy mesophyll leaf structure, which leads to higher values in the NIR channels.

This interaction between leaves and the light that strikes them, often determined by their different responses in the red and NIR portions of reflective light, see Figure 4, is how vegetation is detected using remote sensing. The objective of vegetation analysis from spectral measurements, often, is to reduce the spectral data to a single number that is related to physical characteristics of vegetation (e.g. leaf area, biomass, productivity, photosynthetic activity, or percent cover) , while minimising the effect of internal (e.g. canopy geometry, and leaf and soil properties) and external factors (e.g. sun-target-sensor angles, and atmospheric conditions at the time of image acquisition) on the spectral data.

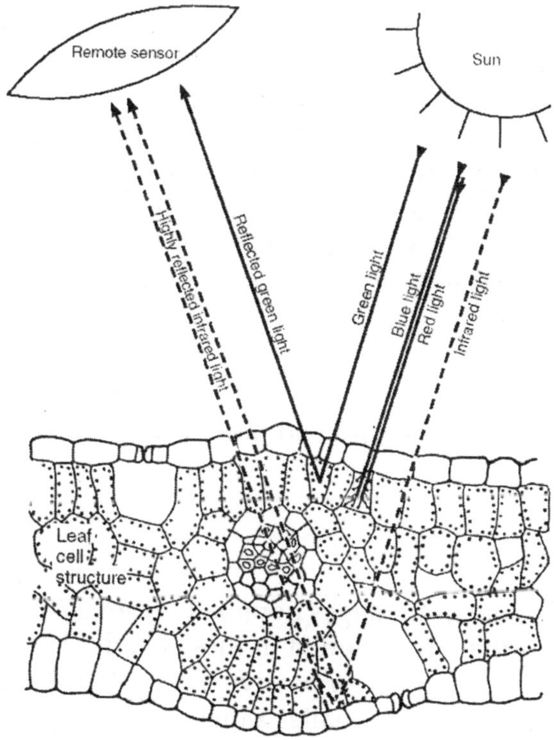

Figure 4. Schematic reflectance of a typical green leaf in cross section.

Vegetation Indices (VI's) were developed in an attempt to obtain this objective from remote sensors by taking advantage of the differences in the reflective responses of vegetation in the red and NIR wavelengths. Although VI's are often hampered by limitations in dealing with the complex nature of real-life vegetation canopy interactions, they have gained widespread popularity due to the benefits of remote sensing's high spatial density and extent, and the value added to generic, rather coarse-scale vegetation modelling. LANDSAT Thematic Mapper ™ and the Système pour l'observation de la Terre (SPOT) sensors, and many other remote sensing instruments, have channels situated in the red and NIR. For example, the red and NIR bands for LANDSAT TM are band 3 (630-690 nm) and band 4 (769-900 nm), respectively. Most vegetation indices are combinations of these two reflective bands. The most common linear combinations are the simple ratio (NIR/Red) and Normalised Difference Vegetation Index (NDVI) = (NIRRed)/(· NIR+Red). Previous research has shown positive correlations exist between foliage presence, including measurements of LAI and plant condition, and vegetation indices such as the simple ratio and NDVI. For a comprehensive listing of vegetation indices refer to Tian, Kaufman and Tanre, Thenkabail et al. and Leprieur et al..

While the amount of leaf is one determinant of the signal strength in the reflective portion of the EMS there are several other important factors which control the acquired value. These include the sun-target-sensor geometry. This will control the amount of shadow contributing to the signal; the shadow may be driven by insolation effects due to regional topography and may also be influenced by vegetation shadowing. This effect, termed the bidirectional reflectance distribution function (BRDF), is characteristic of vegetation structure. Other effects in the reflective portion of the EMS include changes in soil colour and changes in the observed signal due to changes in the atmospheric component of the signal, including atmospheric precipitable water, atmospheric aerosols (e.g., dust), and changes in the response of the sensor over time. Primarily, the reflective portion of the EMS has been used for:

1. identification of rice;
2. area estimation of rice;
3. estimation and prediction of crop yield; and
4. crop damage assessment

Microwave Remote Sensing

The microwave portion of the EMS ranges nominally from 0.75 to 100 centimetres. Radio signals have wavelengths that are included in these bands. These systems can either be active (the sensor sends its own signal) or passive (the background signal from the Earth's surface is observed). There are five smaller range which are used for remote sensing. These are :

P band	100 – 30 cm;
L band	30 – 15 cm;
S band	15 – 7.5 cm;
C band	7.5 – 3.75 cm; and
X band	3.75 – 2.4 cm.

RADAR (RAdio Detection And Ranging), is an active system based upon sending a pulse of microwave energy and then recording the strength, and sometimes polarisation, of the return pulses. The way the signal is returned provides information to determine characteristics of the 12 landscape. RADAR has been used in the determination of near surface soil moisture, and the identification of rice crops based on the presence of standing water. Primarily, the microwave portion of the EMS has been used in agriculture for:

1. identification of rice
2. area estimation of rice; and
3. estimation and prediction of crop yield.

Thermal Remote Sensing

The thermal portion of the EMS ranges nominally from 3.75 to 12.5 micro meters. The radiant energy observed by sensors is emitted by the surface, be it land, ocean or cloud top, and is a function of surface temperature. Models have been developed to allow surface temperature to be extracted from thermal remote sensing. Prata et al. review the algorithms and issues involved in the calculation of land surface temperatures, denoted *Ts*. Thermal remote sensing is an instantaneous observation of the status of the surface energy balance (SEB). The SEB is driven by the net radiation at the surface. During the daytime this is usually dominated by incoming shortwave radiation from the sun, the amount reflected depending on the albedo of the surface. There are also up and down welling longwave components. At the ground surface, the net allwave radiation is balanced between the sensible, latent and ground heat fluxes. Over long periods of time the ground heat flux averages out, and the SEB represents the balance with the sensible and latent heat fluxes. During the day the measured surface temperature at the Earth's surface is, in part, dependent on the relative magnitude of the sensible and latent heat fluxes.

The surface energy balance at any instant is given by:

$$R_n = \lambda E + H + G \quad (1)$$

where:

R_n is net all wavelength radiation (Wm^{-2}); λE is the latent heat flux (of evapotranspiration (ET)) (Wm^{-2}); H is the sensible heat flux (Wm^{-2}), or the energy involved in the movement of the air and its transfer to other objects (such as trees, grass etc); and G is the ground heat flux into the soil or other storages (Wm^{-2}).

λE denotes the amount of energy needed to change a certain volume of water from liquid to vapor, either by transpiration or evaporation. The combination of these two fluxes is called evapotranspiration. The net available energy (*AE*) at the Earth's surface, which is available for conversion to other forms, can be written as:

$$AE = R_n - G = \lambda E + H \quad (2)$$

where the terms are defined as above. A key factor determining the observed surface temperature is the partitioning of the *AE* into the latent and sensible heat fluxes. This is 13 governed by the amount of water available and the ease with which it is transferred from the surface to the atmosphere, via λE.

Thermal remotely sensed data can also be recorded at night. During the night the SEB is dominated by the release of heat from the ground, which was absorbed during the daylight hours. The release of heat during the night is governed by how much was absorbed during the day and, the rate at which it is released after sunset. This is a function of the environment's capacity to store heat, which also depends on the amount of water stored in the environment. The thermal portion of the EMS has been used to determine:

1. surface temperature estimation (including water temperature); and
2. moisture availability (*ma*) mapping.

Integrating Remote Sensing with other Data Types in a Geographic Information System (GIS)

Remotely sensed data (visible, thermal, microwave), GIS data layers (soils, geology), point based measurements (rainfall, soil moisture, biomass) or model outputs (biomass, soil moisture) all have spatial and temporal attributes associated with the data attribute, and can be integrated. Economic situations and social indicators will have a time and may have a space associated with the data attribute and can also be integrated. The integration of several data types will allow factors such as crop types, yield estimates, and water use to be determined more objectively. Many GIS data layers cover entire regions. However, these are often produced from the spatial interpolation of point samples; this is especially the case for some meteorological surfaces. Some physical parameters, such as soil water holding capacity, which are assumed to be time in-variant, only need to be mapped once. Remote sensing provides repeated measurements, at a particular spatial scale and electromagnetic wavelength, which allows dynamic environmental conditions, such as soil moisture and vegetation cover, to be monitored.

The data structure in the GIS can be conceptualised as being pancakes of remotely sensed data (TM or SPOT), which are intersected at right angles by skewers of point based data (Bureau of Meteorology (BOM) rainfall and air temperatures). In the data, two dimensions (the pancakes) are spatial, latitude and longitude, and the third (the skewer) is time. For some issues (e.g., groundwater monitoring) it can be important to include elevation (or depth) as another dimension in the data structure. Both data sets (point based BOM and remotely sensed) often have vastly different spatial and temporal scales. For example, TM data are spatially dense, with a 30 metre spatial resolution, and are recorded over large areas in a 14 matter of seconds at a specific time for specific wavelengths. This means, for the extent of the image, remotely sensed data are a 'census' (high spatial density) at a particular spatial scale recorded at a specific time. Depending on the amount of cloud coverage and the satellite repeat characteristics, optical remotely sensed data may only be available monthly (low temporal density). On the other hand, meteorological data are recorded at specific points, which may be separated by tens to hundreds of kilometres (low spatial density). Meteorological variables are usually recorded daily (high temporal density), as either integrals (e.g., rainfall and wind-run totals), or extremes (e.g., maximum and minimum air temperatures). Thus, remotely sensed data are spatially dense but temporally sparse, while meteorological data are spatially sparse but temporally dense.

Data of varying degrees of spatial and temporal density can be incorporated into a GIS. Spatial and temporal resolutions of the data will vary, depending on, among other things, the issues being addressed. The advantages of using time series of several data sets, including those not observed remotely, is illustrated in the following example from Barrs and Prathapar and Barrs and Prathapar. An irrigation company may want to estimate crop area in order to plan for the marketing of specific crops. To get an early estimate of rice production, a November image was acquired, in which standing water was used to estimate rice area. Standing water can be easily identified in the NIR portion of the EMS where it has a low reflectance (see Figure 3). In order to map other crop types and to delineate standing water areas on which rice was not grown, a March image well into the rice growing season was then acquired. This image was used to refine the end of season yield prediction by 'fine-tuning' the irrigated area estimation from the first image. By determining areas that were still standing water, they could ascertain that these were either failed crops or permanent water bodies. The later image was also used to classify other summer crop types, with varying confidence. Other GIS data, like field delineations and known water bodies, could also have been used to adjust the classification. Errors of omission as well as errors of commission can be corrected, increasing classification accuracy. For example, in a land use map, if 90% of a known field is classified as rice, it makes sense to assume the whole field is rice, thus fixing some errors of omission. However, in a land cover map, knowing, for example, that 10% of the paddock planted with rice did not grow successfully can be valuable information for estimating yields, thus fixing some possible errors of commission. The difference between land use and land cover should be noted as these should not be combined in thematic classes.

REMOTE SENSING IN RICE-BASED AGRICULTURE

A basic understanding of rice physiology is essential to the success of remote sensing applications in rice-based agricultural systems. This knowledge can play a critical role in the planning stages of a remote sensing project (e.g., identifying optimum acquisition dates for the purchase of imagery) as well as in the final stages of analysis (e.g., aiding in the delineation of rice paddies or the estimation of growth stages). The growing cycle of rice can be separated into two stages with respect to most analyses of remotely sensed data: vegetative and reproductive. The vegetative stage includes the part of the growth cycle where the plant develops and grows, starting after sowing and ending when the plants start to reproduce. This stage is characterised by a steady increase in plant height and biomass. The reproductive stage starts when the plant stops growing taller and ends after maturity and includes panicle and grain development. It may be beneficial at times to further split the reproductive stage into two categories: reproductive pre-heading and reproductive postheading. Reproductive pre-heading defines the period from panicle primordia initiation to heading and post-heading refers to the period from heading to maturity. The length of the growth cycle of rice can vary from 3 to 6 months for different varieties, and can also be categorised into two main groups for many remote sensing applications: tropical and temperate. Growth cycles for tropical rice varieties last about 110-120 days, while those of temperate varieties usually last around 140-150 days. However, this duration can vary based on cultivar. For example, short duration varieties have been bred with growth cycles less than 90 days. These differences in growth cycle length are due to differences in vegetative stage duration: the vegetative stage can be anywhere from 40 to 120 days in length.

Irrespective of cultivar, reproductive pre-heading duration is about 23-25 days, while reproductive post-heading duration lasts 30-35 days. During the reproductive stage, plant height and biomass typically remain stable at around 100 cm and 2000 gm^{-2}, respectively. The vertical characteristics of the rice plant also change as the plants grow, with stem inclination decreasing and the leaf angle increasing. For different locations, the timing of the growth cycle of rice varies depending on local climate, management, and cultivar planted. Consequently, the timing of rice for a particular site of interest should be known. Australian rice varieties have changed over the past 40 years with Caloro dominating the 1960's, Calrose in the 1970's to mid 1980's and Amaroo from the mid 1980's into the 1990's. This has been associated with development of long-grain, fragrant and Spanish varieties to meet higher priced markets. Short duration varieties are generally desirable because they can be competitive with weeds, require less pesticides and herbicides, utilise less irrigation water, and allow for double cropping in tropical environments. In a temperate environment like southern NSW, short duration varieties are also desirable because they can 16 allow more leeway for sowing and harvesting between the limitations of cold springs and autumns. Timing of rice in southern NSW is summarised as:

1. placed under permanent flood and aerially sown in late September/early November;
2. canopies emerging during late October/late November;
3. flowering by late January/early February;
4. de-watered in late February/March; and
5. harvested in March/May.

Other summer crops in NSW include corn, sorghum, and soybeans, while winter crops include wheat, barely, oats, and canola. Pasture is grown in both seasons. Citrus, stone fruits, and grapes are also grown in the area. Some of these crops may use remnant soil moisture after a flooded rice crop, whereas others are furrow or drip irrigated. There is an interest in monitoring other crops than rice within the irrigation areas of southern NSW from remote sensing. Of these other crops, there is a good potential for remote monitoring of corn and soybeans. The spectral reflectance characteristics of corn and soybeans along the EMS are slightly different in shape and amplitude allowing for differentiation between these two crops. Multitemporal remote sensing data has been used to estimate soybean and corn crop characteristics such as yield, LAI, biomass, plant height, development stage, and crop proportion. For single date imagery, however, the timing of image acquisition can greatly influence classification results since confusion between spectral signatures can occur due to differences in crop growth stages.

That is, on the day of image acquisition, the two crops could look spectrally similar. Moderately high correlations (from 0.7 to 0.85) have been reported between several soybean and corn crop characteristics when related to VI's. Soybean was correlated to standard NIR and red-based VI's , whereas corn crop characteristics were more highly correlated with VI's that include at least one MIR band. Mature soybean crops have higher reflectance in the NIR and lower reflectance in the red portions of the EMS than corn, resulting in detectably higher standard VI values for soybeans. This means that three of the main summer crops (rice, corn, and soybeans) can potentially be discriminated from each other using remote sensing. However, to do this, more than one image throughout the growing season might be needed in order to take advantage of spectral differences due to the phenology of these crops. Remote sensing based applications, then, will not only take advantage of both the characteristics and timing of growth cycle, but will also consider the spectral reflectance of different crops. Since the basic spectral patterns of rice must be understood. The reflectance from rice, like all green vegetation, can be summarised by a generalised vegetation response as seen in Figure 3. It is the differences in this basic vegetation response that allow discrimination between vegetation types. However, these vegetative-type responses are harder to differentiate between each other than a non vegetative-type response like soil or water. This is true since non vegetative-type features usually reveal drastically different response curves when compared to vegetation (Figure 3). As irrigated rice fields are flooded, the spectral characteristics of water can be used to distinguish potential rice paddocks and provide an early estimate of rice area. Inaccuracies result, however, when this early estimate is not adjusted by a later image, which can aid in elimination of permanent water bodies and other irrigated crops from the classification. The visible and near infrared wavelength response of rice, once the vegetation starts to cover the water in flooded paddocks, is much the same as other crops. However, rice was found to be more distinguishable from other crops due to its water absorption characteristics by including middle infrared (MIR) wavelengths in the crop discrimination.

It is also very important to understand the interaction of the reflectance of vegetation in key bandwidths of the EMS to relevant vegetation characteristics such as biomass or leaf area index (LAI). Since most studies relating vegetation to biomass or LAI use VI's of NIR and red EM portions, these interactions, specifically should be understood. NIR reflectance of rice is directly related to green biomass. It continues to increase from early tillering, where NIR reflectance is about 15%, to heading where it reaches a maximum of about 50%. Post-heading NIR reflectance decreases to about 33% as rice crop green biomass decreases due to death and loss of leaves. Red reflectance of rice is inversely related to green biomass, where it decreases from about 10% at emergence to 2% at flowering, and then increases to about 18% at maturity due to senescence. Some VI's perform better than others, but in general, these interactions result in better correlations between VI's and biomass in the vegetative and pre-heading stages than in the post-heading stage, where VI's tend to saturate.

The saturation of VI's results in a notable limitation to the usefulness of remote sensing in estimating crop biomass, LAI, and, therefore, yields in rice-based irrigated agricultural systems. This saturation is mainly due to the inability to detect the accumulation of biomass that takes place well after crop canopy cover reaches 100%. After canopy closure, NIR reflectance continues to increase significantly, while red reflectance may only change slightly, resulting in minimal changes in a VI. Since NIR is typically the numerator of a VI ratio, and red is the denominator, it takes an inordinately large increase in NIR to increase the overall VI after heading. This is demonstrated in the saturation of NDVI above LAI's of 2.5 to 3 and fraction of intercepted photosynthetically active radiation (*fPAR*) above 94%. In irrigated agriculture, LAI's often exceed 3, which decreases the accuracy of yield measurements from remote sensing. However, this is somewhat offset by the fact that growth stage information received from remote sensing is often useful and that microwave remote sensing measurements can provide structural measurements after LAI's exceed 3. The applications of remote sensing relevant to this discussion can be separated into 6 main categories as determined by the bulk of the current rice-based remote sensing literature. These are:

1. crop type identification;
2. crop area measurement;
3. crop yield;
4. crop damage;
5. water use/ moisture availability (*ma*) mapping; and
6. water use efficiency.

Crop Type Identification

The most commonly practiced application in remote sensing of agriculture is mapping land cover to identify crop types. This process primarily uses the spectral information provided in the remotely sensed data to discriminate between perceived groupings of vegetative cover on the ground. The spatial and temporal information included in single date and time series data, respectively, usually play a secondary role, but can also aid in the classification procedure. Discrimination of crops is usually performed with 'supervised' or 'unsupervised' classifiers. The basic difference between these types of classification is the process by which the spectral characteristics of the different groupings are defined. Common clustering algorithms include maximum likelihood, minimum distance to mean, and parallel piped. Unsupervised classification relies upon a computer algorithm to define natural groupings of the spectral properties of the pixels in an image. After initial classification, the analyst attempts to assign the perceived groupings *a posteriori* (after the fact). Problems arising from using this methodology are related and include classification of meaningless groupings, idiosyncratic definition of initial number of classes, and subjectivity involved in combination of similar classes. However, if land cover classes are spectrally well separated, adequate results may be obtained.

Because the unsupervised classification algorithm will automatically output the often arbitrarily defined number of classes input to it, the resulting classes are frequently nonrepresentative or meaningless. That is, the initial classification can be drastically different based on the number of classes the analyst originally requests. Some of the classes generated may have little meaning with respect to reality because they contain more than one functional or 'on the ground' class. Others are too specific (i.e., several spectral classes form one functional class) and need to be recombined. Spectral classes that contain more than one functional class are frequently harder to deal with because the analyst will need to reprocess them. The recombination of overly specific classes can also be problematic since the process is often subjective. This subjectivity is regularly expressed by results that are not easily duplicated and non-standardised (often represented by abrupt land cover changes at management boundaries). The advantage of an unsupervised approach, however, is that no *a priori* (before the fact) knowledge is needed. This can be beneficial when every crop type in a study site may not be known, or when attempting to discriminate between groups of crop vigour (Barrs and Prathapar, 1994). Unsupervised classification also avoids problems with the user biasing the classification with improper or poorly represented training data, which can be the case in supervised classification.

Supervised classification, which is the most common classification method in agricultural areas, requires *a priori* knowledge. It relies on the analyst to define the perceived groupings by identifying homogeneous areas, called training sites, from either *in situ* collection or directly from the image. These training sites are statistically analysed and then used to assign every pixel in the image to the group in which it has been determined a member. Problems 19 may arise when defined training sites are a poor representation of the group's spectra. For example, training data with multiple modes in the training class histogram suggests that there are at least two different types of land cover within the training area. Also, positive spatial autocorrelation exists among pixels that are contiguous or close together (i.e., adjacent pixels have a high probability of having similar brightness values), which can cause a reduction in variance between adjacent pixels. This reduction in variance can make large clumps of contiguous training pixels less representative of a particular cover type over the extent of an entire image and is why several single-pixel training sites located spatially apart from each other can result in better classifications than large clumps of contiguous training pixels. Spatial autocorrelation also means that pixels in a remotely sensed image should not be thought of as entirely discrete features independent of their neighbours, but rather a set of continuos features influenced by their neighbours. Therefore, care must be taken to collect representative and non-autocorrelated training data. When such care is taken, classification for crop identification can be quite effective.

In general, classification accuracy assessment involves sampling classes of the thematic map to summarise the amount of overlap between the classified map and what is present on the ground, although the performance of the classifier can also provide information on map accuracy. The 'ground truth' information is usually gathered from existing maps, aerial photography, or field surveys using any number of sampling schemes (e.g., simple random sample, stratified random sample, or systematic sample to name a few). However, because of differences in the accuracy of the 'ground truth' source and the sampling scheme used to gather the information, a certain amount of ambiguity is present in any classification accuracy summarisation. Certainly, the resulting accuracy assessment will be influenced by the reliability of the 'ground truth' source. Validation data might be less reliable than the classification itself, yet is used as 'ground truth' because it has been traditionally the standard by which management decisions have been made. For example, agricultural census' of large areas in rural-based countries may be extremely generalised, yet are often used to 'ground truth' a classification of fine spatial scale (e.g., 30 metre) satellite imagery. In this case, the classified map is probably more accurate than the 'ground truth' making unclear as to what the discrepancies between the classified map and the 'ground truth' are attributed to.

Another source of confusion introduced by classification accuracies is the way in which the accuracies are reported. It is advisable to differentiate between what is loosely known as producer's accuracy and user's accuracy. Producer's accuracy, here, is a strict comparison of the number correctly classified on the map versus the number of that same class determined by the 'ground truth' (i.e., per cent correct). This measurement only considers errors of omission ('ground truth' points that were left out in the classification). User's accuracy, on the other hand, considers both errors of omission and commission.

To achieve high classification accuracy, multitemporal, multi-sensor data is often integrated into the classification procedure. However, incorporating these data into a classification does not necessarily improve all crop type accuracies. For example, Barbosa et al. attained higher accuracies for several crop types using single date imagery as opposed to multitemporal imagery. These results depended on both the timing of the single date image as well as the accuracy assessment methodology used. For example, rice crop identification using one single date spring image produced equivalent or higher accuracy than the multitemporal classification. This specific example could be related to the ease by which the water in flooded rice fields can be separated from other areas at the time of image acquisition. In contrast, some authors believe that the early (spring) estimate of rice crops can be improved by use of a later image, which can aid in elimination of permanent water bodies and other irrigated crops from the classification. The 'law of diminishing returns' is most likely at work in this situation; slight increases in the accuracy of rice crop identification may not be worth the cost of purchasing another remotely sensed image unless the impact of the assessment on management is relatively high or potentially costly.

The performances of different sensors for crop identification have been tested over varied geographic areas and crop types. These most commonly include broadband optical (e.g., LANDSAT, SPOT, and AVHRR), and microwave (e.g., ERS, RADARSAT, JERS) used alone or in combination in the form of either single or multiple date imagery. In general, multitemporal microwave imagery results in roughly equal classification accuracies as single date optical imagery when specifically considering rice (i.e., about 90%) or other crops. Because of the complementarity of the data, the fusion of both optical and microwave imagery has resulted in higher overall and individual crop classification accuracies (> 95%) than either produce alone. Higher crop classification accuracies have also been achieved by the combination of GIS and remotely sensed data. This method of classification depends on accurate GIS data and is sensitive to missing boundary lines. By using GIS data, Aplin achieved greater classification accuracies in 'constrained' areas having well-defined boundaries (e.g., agricultural fields and urban areas). Conversely, the inclusion of GIS data decreased classification accuracy in 'unconstrained' areas or those areas having poorly defined boundaries (e.g., grassland and bare soil areas). Similarly, others have improved classification accuracy using the contextual or landscape information inherent in the imagery. These techniques require no *a priori* knowledge, and therefore, are more flexible because they do not rely on an updated and accurate GIS layer. However, the user has more control when using GIS data, which can result in very accurate results. Finally, hyperspectral remote sensing allows for the classification of imagery using many narrow bands along the EMS. These narrow bandwidths permit the use of very specific spectral characteristics for optimising the categorisation of agricultural properties such as biomass and leaf area index as well as classification of crops and crop stress. The narrow bands in hyperspectral data are often ordered sequentially along a wide bandwidth (e.g., every 10 nm from 470 – 2500 nm) providing continuos spectral information over that EM portion. This continuos nature of hyperspectral data can provide very valuable information that can increase classification accuracy and is why this data source will be frequently used in the future. However, the tremendous spectral information comes at a cost; the datasets are much larger than broadband datasets and thus are harder to maintain in terms of data management and processing. Also, in multi-temporal analysis, accurate atmospheric correction of the hyperspectral data will be more critical than for broadband data.

Crop Area Measurement

Crop area measurement is a very common practice in agriculture. Remote sensing is often used for this purpose because of its strengths in regard to spatial extent, temporal density, relative low costs, and potential for rapid assessment of spatial features. Many of the same issues concerning crop type identification also affect crop area measurement from remotely sensed data. This is because crop type identification is a necessary first step to area estimation. In many cases, though, crop type identification is more concerned with classifying all crop types from each other, where area estimation often is concerned with only a few target crops. In either case, these two applications are frequently performed in sequence: first crop identification and then area estimation. There are a few issues that are not exclusively related, but tend to more specifically pertain to crop area estimation, including positional accuracy, mixed pixels and pixel size, and a mismatch between individual and overall accuracies of the results. Positional accuracy, here, can be defined as the difference in the position of a feature on a map compared to the feature's real world or 'true' position. As such, the position of boundary lines on the map, for instance, are most likely not where they are in the real world, but are more accurately represented as a belt or swath around that boundary line on the map. This swath contains the 'true boundary line' and has a width that is inversely related to the scale of the source. For example, as the scale of the source gets smaller (area representation gets larger) the width of the swath around the line generally represents a larger distance, making the positional accuracy decrease.

In other words, as the swath gets wider, the relative certainty of the position of the 'real boundary' gets smaller and the error likely increases. Therefore, when attempting to estimate crop areas accurately, considerable thought should be given to achieving high positional accuracy of crop boundary lines if GIS data are used in conjunction with remote sensing. Ground validation points are extremely important, in this case, to quantify positional accuracy and thus spatial uncertainty. Knowledge of spatial uncertainty is essential in making appropriate managerial decisions on crop area measurements. For example, if crop area estimation is performed in an attempt to monitor year-to-year land use, it is extremely important to know how much potential error is included in the estimate solely due to uncertainties. With this knowledge, the manager has a better idea if the year-to-22 year difference in area is reliably measured from the remotely sensed data or if it is 'absorbed' by the inaccuracies of the dataset itself. Pixel size of the remotely sensed data also affects positional accuracy of boundary lines in crop area estimates, and therefore should be considered for its appropriateness to a particular application. One prominent issue is the relation of the pixel size to the paddock size (or feature element) being measured. Although for large areas AVHRR data is very attractive due to its spatial extent and high repeat cycle, the 1 km pixel size is often disproportionately larger than paddock sizes. This may not be an unsurmountable problem when monitoring areas with many contiguous smaller fields of the same crop type, but at the same time, must be considered. In an attempt to either compensate for mismatched pixel-to-field sizes or to increase the accuracy of area estimations, the spectral characteristics of impure or 'mixed' pixels can be 'unmixed' by linear mixture modelling.

In linear mixture modelling, the analyst assumes that any mixed pixel's spectral signature is made up of a combination of pure spectral signatures of all the separate land cover types contained in that pixel in proportion to the area of which that cover type is found in the pixel. The spectral signatures resulting from a pixel containing only one cover type (pure pixel) are known as end members. Proper unmixing of mixed pixels relies upon the identification of good end member spectra. Different combinations and proportions of the end members are combined to best match the signature of the mixed pixel, supposedly allowing for a better estimate of crop areas. Problems can arise when end members are either not a good representation of a particular land cover class, or when a land cover class' end member is not collected. Even when numerous and representative end members are gathered, results can still be spurious due to confusion caused by mixed pixel spectra being explained by multiple possibilities of end member combinations. Also, green crops can produce very similar end members at certain times of the year, resulting in erroneous unmixing in agricultural systems. As with identification of crop types, ground validation is also very important in area estimation. Traditional estimates of crop area have come from census data over rather large areas, and are usually rather gross and usually contain little specific spatial context. This means that there is often no explicit locational knowledge contained in the 'ground truth'. This leads to very ambiguous results where overall values may be closely related, but the analyst often has no idea what sort of spatial variation exists between the 'ground truth' and the estimates from the remotely sensed data, leaving the comparison less meaningful. It is important, then, to make sure that the 'ground truth' data is defensibly more reliable than the observation as well as spatially explicit, wherever possible.

Often, there exists in the validation of area estimations, a mismatch between reported overall and individual accuracies. Many authors represent areal accuracy as an overall summation of several paddocks or districts. This tends to reveal extremely accurate results, often exceeding 99%. However, this can be misleading because there often exists large areal errors in the individual paddocks or districts that were added together to generate this statistic. Unfortunately, this mismatch between overall and individual areal accuracy is rarely discussed, revealing a perceived dichotomy in the results. The discrepancy between the overall accuracy and individual district- or paddock-level accuracy is due to errors of underestimation and overestimation cancelling each other out. That is, when several areas are added together, the overestimates are often almost exactly offset by the underestimates, resulting in a good overall estimation of area even though individual error is much larger. In a study at the CIA in southern NSW, this individual areal error was seen to drop considerably when just two areas were added together, and continued to approach zero as more areas were summed. This phenomenon can be exploited for management purposes, but care must be taken to ensure that proper interpretation of areal accuracy is achieved for specific management situations. That is, it is only appropriate to use the summed or overall areal accuracy when the overall area is the specific concern.

Crop Yield

Crop yield forecasts can greatly influence farm-level management decisions, such as fertiliser applications and water delivery, as well as provide a means for farm income assessment. Consequently, individual farmers and district-level land managers show great interest in producing rapid and accurate estimates of crop yield, both locally and regionally. In the past, the standard yield estimation procedure included the analysis of crop cuttings at randomly sampled ground plots during harvest, or meteorological regression models using rainfall and past yield data. These methods often produce results that are either not timely nor spatially explicit. Though still used, these methods are being replaced by estimation of crop yields using remote sensing because of its ability to produce results quickly and spatially. Using this technology, it was found that spatially meaningful estimates of yield can be made as early as 1 to 3 months prior to harvest, thus impacting management reaction time to yield forecasts. Positive correlations exist between measurements of LAI, plant condition, and VI's. Based on this relationship, and assuming that LAI (or biomass) is related to yield, there are 3 main approaches used to forecast yield. These are: (1) correlate yield with NDVI; (2) correlate yield with integration under the NDVI curve, denoted ∫NDVI; or (3) simulate yield with crop models (sometimes using remotely sensed inputs). For a detailed review of crop yield modelling using remote sensing.

Both regressions of VI and ∫NDVI studies generally attempt to estimate yield using remotely sensed data alone. A direct comparison of VI response to yield can result in highly correlated regression equations. For example, Harrison et al. achieved correlation coefficients ® as high as 0.9 for calrose rice and 0.79 for Inga at the CIA using the greenness ratio of MSS7/MSS5. This study developed the yield model based on imagery acquired when the 24 crop was at booting stage, the time when rice displays green leafy surfaces. Promising correlation coefficients were also shown between reflective bands of TM imagery and harvested grain yield (up to 0.92 for band 7). These relationships can be used to estimate rice yield at district levels with reasonable success. For example, Patel et al. was able to estimate district rice yield with an accuracy ranging from 86% to 98%.

Likewise, ∫NDVI studies have produced similar results with relation to rice yield. Rice yields have been estimated within 3% of official values for 3 consecutive years in one study, while correlation coefficients have been reported as high as 0.93 for millet yield in another. However, the ∫NDVI method is highly sensitive to remotely sensed data availability, which can result in gross yield underestimates when portions of the growing season are missing. Until just recently, the only operational satellite sensor with a repeat cycle sufficient for ∫NDVI studies of crop yield has been NOAA AVHRR. The most notable additions to this list are the TERRA (ASTER and MODIS sensors), and Ikonos satellites both launched in 1999. Anticipated sensors could also impact the usefulness of ∫NDVI studies by increasing the number of satellites with relevant repeat cycles. These include Quickbird-1, and Orbview 3 and 4. The definition of the proper time interval for ∫NDVI studies has been disputed, but shown to only be significant when integrated over the reproductive stage (for millet). For rice, where LAI usually exceeds 5, VI's saturate, limiting the usefulness of this approach, and suggesting that different methods might achieve better results. To date, the ∫NDVI method has mostly been used in dryland farming. In an attempt to achieve highly accurate yield estimates, many models have been developed which aim to simulate plant growth. Such plant growth models need to be validated to ensure that they meet a desired level of predictive capability. Plant growth models can be validated by comparing simulated and *in situ* measurements of vegetation parameters (e.g., biomass or yield). However, plant growth models may also be validated by comparing simulated reflectance with that measured by remote sensing. This can be achieved by inverting either empirically-based relationships, or physically-based radiative transfer models. This means that a plant growth model estimate of cover can be converted, through a radiative transfer model, to an estimate of reflectance, or recalibrated through the estimation of key crop variables like LAI or above ground biomass. This recalibration of crop variables links well to simple growth models because they can be directly related to *fPAR* , a key variable in most of these models).

Because the radiative transfer model approach can be complicated and necessitates many input variables, the recalibration of simple models is often preferable. Recalibration is also, very often, a better method for yield estimation than either using the remotely sensed data as direct input to simulation models, or ∫NDVI because it is not as limited by the availability of remote sensing data. The recalibration approach using remote sensing has also been shown to drastically improve non-calibrated simulation models.

Crop Damage

Of great concern in agricultural systems is the loss of productivity due to crop damage and its negative impact on meeting the increasing demands for food globally. Rice damage and subsequent yield reduction can occur for many reasons including, various pathogens, insects, and weeds and often possesses a complex interaction with cropping practices. Cropping practices such as differences in water depth and fertiliser applications have been shown to affect crop production. Likewise, irrigation with wastewater can cause heavy metal contamination in soils, also resulting in crop damage and yield reduction. Natural variables, such as climate, can also cause damage from either seasonal temperature or precipitation (drought) variations, although irrigated systems are generally sensitive to cold temperature events only. In southern NSW, depth of ponded water is used to regulate plant temperature. The rice panicle is kept under water, where during the night it remains warmer than the surrounding air temperature. Rice leaf blast epidemics may be related to temperature as well, leading to model simulations of yield losses due to current global warming trends. Crop damage research using remote sensing can be simplified into two different categories: those that measure the source of damage (direct measurement), and those that measure the effect of the source of damage (indirect measurement). In certain situations, the source of crop damage can be measured directly from remote sensing. These studies include assessment of salinisation, weed infestations, and waterlogging.

Saline soils have been shown to have higher reflectance in both the visible and NIR wavelengths than non-saline soils from both ground and satellite radiometric measurements allowing for mapping of salinisation. It should be noted, though, that successful mapping of salinisation results only when salt scalds are clearly visible on the soil surface. This is not the case for all salt affected soils, which means that salinisation is not always easily mapped from remote sensing. Detection of water from dry land is uncomplicated spectrally leading to the classification of potentially waterlogged areas. However, these areas should be integrated with an accurate land use map as permanent water bodies and natural wetlands could quite easily be included in the classification otherwise. Ground-based electromagnetic (EM-31) surveys and airborne radiometrics have also proven useful for mapping soil attributes. Weed infestations can also be directly measured with remote sensing if the weeds are prevalent prior to crop emergence. In this case, the living vegetation (weeds) is easily identifiable from the background soil. These infestations are also detectable after crop emergence if their spectra are significantly different from the background soil and crop cover. It was found that weed detection requires high spatial resolution and is sensitive to spatial registration and weed density, which can all affect the accuracy of the results. This spatial resolution requirement limits the usefulness of most current satellite platforms for weed detection and is why most of these projects make use of airborne systems.

In most cases, however, it is impossible to measure the source of the damage directly using remote sensing. For example, the measurement of ducks or insects in a rice paddock is not feasible, whereas, the effects of these pests can potentially be detected through loss of above ground biomass or yield. In this sense, crop damage is almost always associated with crop yield. This means that any of the multiple techniques that have been developed for estimating yield from remotely sensed data, can then be associated with losses due to various forms of crop damage. In specific cases, however, the indirect measurement is not necessarily associated with monitoring yield or biomass losses from a particular event (e.g., an insect infestation), but rather is an attempt to detect possible plant stress over a less discrete period of time. It is well established that physiological stress in crops is related to incident reflectance from crop canopies and that this altered reflectance can be detected with remote sensing. These changes in reflectance are prominent in both the visible and NIR regions of the EMS. An increase in leaf reflectivity in the visible light portion as a response to stress is caused by the metabolic sensitivity of chlorophyll resulting in less efficient absorption of light (increased reflection). NIR reflectance of leaves is more variable at the onset of stress, but has been shown to increase steadily as stress continues in the form of dehydration. NIR also looses distinctiveness in its spectral signature, especially in the water absorption regions, as the curve flattens. These individual leaf responses are often seen at the crop level as losses in leaf area or foliar density in conjunction with increases in shadow and nonfoliar surfaces if the stress continues long enough. Stresses due to dehydration will not generally be a problem in the irrigated areas of NSW, but will affect the surrounding dry-land agricultural systems. Also, the irrigated crops should have similar physiological/spectral responses to other stresses (e.g., disease). In either case, the major objective of such crop damage detection is the link to management decisions in a timely and cost effective manner in order to operate agricultural systems more efficiently.

This is why so much of the current crop damage literature concentrates on precision agriculture. Precision agriculture is a process that heavily relies on observations describing within-paddock variability in order to implement dynamic management decisions that may result in higher crop production. Another key component of precision agriculture is its reliance on technology such as Global Positioning Systems (GPS) and Variable Application Technology (VAT). These technologies, along with remote sensing, make the implementation of the procedure possible. Critical spatial input to precision agriculture decision support systems include crop yield and soil properties. Several studies have shown how to generate these spatially variable inputs through analysis of ground samples, however, most prove to be both time and cost consuming, especially over large areas. These limitations have made remote sensing an attractive alternate source of this spatially variable observational data.

In general terms, the main benefits to the use of remote sensing in precision agriculture are associated with its strengths in mapping both time and space. Seasonal (temporal) and spatial variability of environmental characteristics, mainly crop and soil conditions, can be mapped efficiently with remote sensing. These data can then be used to implement timely management actions as the crops develop. Current limitations to the use of remote sensing in precision agriculture are mainly related to technology. Current satellite systems have fixed spectral bands, too coarse spatial resolutions, inadequate repeat cycles, and long delivery times. Airborne sensors overcome some of the spectral and spatial resolution problems associated with satellites, but these systems generate additional problems with radiometric and geometric correction making it hard to monitor large areas as well as adding difficulty and time to image processing. Both systems (satellite and airborne) also tend to accumulate high costs associated with the large time-series of inter annual data that has to be acquired. Moran et al. discuss in detail the potential role of remote sensing in precision agriculture.

Benefits of precision agriculture not necessarily associated with remote sensing are increased efficiency, reduced risk of environmental hazards, and improved process control. The limitations include cost and complexity of the system, training of users, and delivery of appropriate input data in a timely manner. As such, the potential usefulness of remote sensing in crop damage management, specifically for implementation in precision agriculture is high. However, the limitations of precision agriculture currently make it nonoperational at a regional scale. As seen from various paddock-level studies, the process has merit, but is not yet routinely viable for most farmers or land managers. This is likely to change in the near future as some of these limitations disappear due to advances in technology and research methodologies.

Water Use and m_a Mapping

Knowing crop water use, both temporally and spatially, in irrigated areas allows water delivery to match agricultural demands. Crop water use can be determined either by crop specific empirical models or use of process based models. Both of these approaches require access to ground based meteorological data, usually with a daily time step. To perform either modelling approach over large irrigation areas will require access to a network of meteorological stations with a suitable spatial density and extent to characterise the spatial variation in observed meteorological variables. A commonly used method to estimate crop water use is application of the Food and Agriculture Organisation (FAO) Guidelines for prediction of crop water use. Class A Pan evaporation can be measured if such facilities exist and are properly maintained, or can be estimated from commonly observed meteorological data. The FAO method requires modification of crop coefficients; it does not require remotely sensed data. However, knowledge of different land uses are required to extend this approach spatially. Remote sensing is seen to be a timely and cost effective way to provide these maps to GIS models annually. A derivate of this method have been used by Kirk et al. to provide estimates of crop water use for farm-level irrigation water use efficiency for irrigation areas in South Australia. The location of paddocks was obtained from a GIS data base of paddock level cropping. Many other water balance/plant growth models have been developed which require meteorological data, and soil and generic plant parameters to successfully run. These models can range from empirical to process based, however, as with the FAO approach, suitable maps of land use is required to take these estimates into the spatial domain. Given the issue of advection in irrigated environments the areal weighted up-scaling approach using GIS strata may introduce large errors in the regional crop water use calculation.

Regional remotely sensed *Ts* observations can be used in a resistance energy balance model to provide estimates of the *H*. This allows the λE to be estimated as the residual, obtained by rearranging Equ'n 1. REBMs require effective models for the resistances and other terms (including *Rn* and *G*). Kustas and Norman review the use of remote sensing to estimate λE. REBMs describe the fluxes between soil and plant canopies and the atmosphere in terms of 'resistances'. The models are 'closed' by specified meteorological inputs defined at a reference height above the surface and some assumptions or models defined at or below the earth's surface. To solve a REBM at the time Ts is observed, several surface and near surface meteorological variables (either measured or estimated) are required. These include air temperature (T_a), relative humidity (h) or vapor pressure (e_a), solar radiation (R_s) and wind speed (u) at some reference height above the surface and over a time limit of sufficient length for equilibrium to be assumed. The feasibility and effectiveness of providing such data when only standard daily meteorological data are available has been demonstrated. REBMs also need surface information such as albedo and LAI (or percent vegetation cover) which may be obtained either from reflective remotely sensed data or *in situ* measurements.

During detailed experiments, when specific time-of-day meteorological data are collected, progress in REBM parameter formulation has been made. However, such experiments are limited to short times for small areas. For operational regional irrigation water supply management, obtaining REBM estimates of fluxes at isolated points is of minimal use. Either model inputs or model outputs need to be spatially interpolated to the entire region of interest. The issue of 'interpolate then calculate (IC)' or 'calculate then interpolate (CI)' has received attention to spatially estimate moisture deficit, moisture availability, global solar radiation, and soil properties. To avoid running a REBM at every point in the agricultural system, a number of methods have been developed. The methods are based on the link of the energy balance and water use. For example, leaf temperature can increase when the leaf is green, as stomatal closure to minimise water loss by transpiration results in a decreased λE. At the same time, due to the requirement of the energy balance, H usually increases. Idso *et al.*, Jackson and Jackson *et al.* pioneered methods using daytime T_s for assessing crop health and establishing irrigation scheduling at the field scale. These techniques, culminated in the development of the Crop Water Stress Index (CWSI) which has the form CWSI = 1 − m_{ad}, where the subscript d means the daily integral. The Normalised Difference Temperature Index (NDTI) directly maps m_a regionally. The high spatial density present in remotely sensed data is exploited by using this data as a covariate to spatially interpolate the NDTI in a CI approach.

A number of approaches based on the negative correlation between T_s and NDVI (denoted T_s-NDVI) to provide a measure of environmental stress have developed. Goetz reported that the negative correlation between T_s and NDVI, observed at several scales (25 m^2 to 1.2 km^2), was largely due to changes in vegetation cover and soil moisture. For complete canopies, the slope of the T_s versus NDVI plot has been related to canopy resistance. Nemani et al. found the slope of T_s versus NDVI plot to be negatively correlated to a crop-moisture index. These empirical relationships defined from the slope of T_s-NDVI plots need to be acquired over a range of NDVI and T_s conditions, to allow the 'warm edge' to be calculated with any confidence. It also should be noted that the slope of T_s-NDVI plots is empirically related to m_a. For two different images, with different meteorological conditions, resulting in different atmospheric resistances, the relationship between the slope of T_s-NDVI plot and m_a can be non-unique.

Advances are currently being made that use the T_s versus NDVI plot combined with process based understanding to provide a more mechanistic interpretation of the remotely sensed data. There are two methods currently being put forward. The first, a progression of the slope of the T_s versus NDVI plot approach, describes the data as falling into a 'triangle'. The second, the Vegetation Index/Temperature Trapezoid (VITT) promotes the idea of data falling into a trapezoid. The VITT is an evolution of the CWSI. The unifying feature for both the 'triangle' and the 'trapezoid' is the appearance of the 'warm edge', the line between the dry bare soil point and the fully vegetated point (for the trapezoid its the fully vegetated stressed point). For any point, if both the vegetation cover and T_s can be measured, then m_a can be empirically defined for that image. The top side of the 'trapezoid' may collapse, forming a 'triangle', as a function of the remotely sensed spatial scale and the timing of acquisition during the growth cycle. The main limiting factor for use of thermal remote sensing for operational management of irrigated areas is the repeat characteristics of remotely sensed data with suitable spatial resolution. Another difficultly in many irrigated rice growing areas is the measurement and specification of lateral energy flow (advection) from surrounding non-irrigated areas, which is usually not represented in REBMs. For example, in an advective environment, Dunin states "the discrepancy (between lysimeter and Bowen ratio values of λE_a) reached 80 Wm^{-2} by noon and persisted in excess of 100 Wm^{-2} throughout the afternoon". This illustrates the difficulty in estimating λE_a using models that don't include components for lateral flow of energy (as in most REBMs).

Water Use Efficiency

With population growth comes a decrease in living space (or global land area per capita) and thus, increased competition for land and water resources. These higher demands on water and land, among other things, result in a need for more efficient use of the resources. Assessing and improving Water Use Efficiency (WUE) in agricultural systems, then, will become exceedingly important as the demand for food production on these limited resources continues to increase. Water savings from these systems, in particular, could affect the regional and global water balances as the area of land placed under agriculture is expected to increase considerably in the near future. WUE of the irrigated agricultural lands of southern NSW has become increasingly relevant, as the Snowy Water Inquiry into the environmental issues associated with the corporatisation of the Snowy Mountains Hydro-electric Scheme was initiated in 1998. Ninety-nine percent of the runoff from the Snowy Mountains is diverted inland to generate hydroelectric power, and irrigation water for agriculture. The Scheme provides an average of 1200 Gl per year to the Murray River and 1210 Gl per year to the Murrumbidgee. This water is used in the production of approximately A$1.5 billion per year worth of irrigated agricultural products.

Benefits of the scheme include electricity, increased urban water supply, recreation, dilution of salinity in the Murray River, stock watering, displacement of an estimated 5 million tonnes per year of carbon dioxide emissions to the atmosphere (the amount that would be produced if the same quantity of electricity was generated by coal-fired thermal power stations), as well as providing the water for, among other crops, a 1.5 million tonnes of yield per year rice industry. Environmental impacts from the changes in flow and discharge regimes across these rivers include reduction in channel size, increases in groundwater levels, salinisation, erosion of river banks, changes in wildlife habitat, weed invasion, and a reduction of water quality. The Snowy Water Inquiry was established by the Commonwealth, New South Wales and Victorian governments to determine how to best balance the environmental, electricity and irrigation interests of the region. Environmental issues of the inquiry included impacts on rivers, river channels, ecology, the lower Snowy River, and environmental flows. Economic issues included impacts on electricity generation, impacts on agriculture, and impacts on recreation and tourism. Social issues include the alteration of social values of heritage, culture, and community.

The final report delivered on 23 October 1998 presented 23 options for resolution, of which, composite option D was supported. Option D involves increasing the flow of the Snowy River below Jindabyne Dam to 15% of the pre-Scheme mean discharge . This option was chosen because it would provide environmental gains at minimal cost to agricultural, hydroelectric, and different government agencies. Whether the environmental gains of this option would be significant, or cost to agriculture minimal has been questioned. Also, it is questionable whether reducing one environmental problem is worth adding to another as increasing environmental flows in the Snowy River will translate into greater greenhouse emissions from subsequent thermal-powered electricity generation. The outcome of this inquiry will, no doubt, require irrigation managers to be more efficient with the water that is supplied to them. This increased WUE may occur through a combination of technology and changing management practices, and will most likely necessitate an analysis of WUE at different scales. Regional-scale analysis of WUE would be most helpful in defining gross efficiency relationships and help target areas that could potentially benefit from changes in management practice. With this analysis, the baseline WUE could be determined and possibly whether or not plant breeding since the 1980's or 1990's has influenced WUE on a regional level. As such, regional WUE will be the focus of the following discussion.

Stanhill defined water use efficiency (WUE) both hydrologically and physiologically. Hydrological WUE is the ratio of evapotranspiration to the water potentially available for plant growth. It is expressed as a unitless percentage or fraction (0–1). Physiological WUE is a measure of the amount of plant growth for a given volume of water. The definition of physiological WUE can be defined for different measures of 'plant growth' and 'volume of water'. Turner notes that care is needed when defining WUE. For example, in some studies 'plant growth' has been measured in units of net biomass (including roots). In various previous studies the units of 'volume of water' is either the total transpiration, total evapotranspiration, total water input, or the amount of precipitation plus initial soil water at the time of sowing.

Sinclair et al. introduced different time scales for several definitions of WUE. The temporal scales ranged from an instant, through daily, to a growing season. Intrinsically linked to a range in temporal scale is a range of spatial scale (Table 2), which can extend from a single leaf, through canopy, to field, farm level, and regional assessment. The scales are linked: for example, leaf WUE (in the order of 10's of cm^2) will usually be measured over a short time (e.g., a second to daily). On the other hand, farm-level and regional WUE (in the order of 10's to 1000's of km^2) will usually be measured over a longer time (e.g., a growing season). To date, there has been very little research focussing on farm level or regional assessment of WUE.

Table 2: Different Spatial Scales of Physiological Wue, Corresponding Temporal Scales, And Some Common Units In Which They Are Measured.

Spatial scale	Temporal scale	Units
single leaf (10's of cm^2)	second – day – growth stage	$mgCO_2/gH_2O$ or μmol/mmol
canopy (1 to 10's of m^2)	second – day – growth stage	$mgCO_2/gH_2O$ or μmol/mmol
field (100's of m^2 to 10's of ha)	day – growth stage – season	mg/gH_2O or kg/ha.mm
farm (1 to 100's of ha)	growth stage – season – year	kg/ha.mm or tonne/km^2.TL
region (10's to 1000's of km^2)	season – year	kg/ha.mm or tonne/km^2.TL

Single-leaf WUE is commonly defined as the net CO_2 uptake per unit of transpiration. On a continuous basis – that is, at any instant within a day – it is expressed as the ratio of leaf net photosynthetic rate to leaf transpiration rate, or at the daily time-step it is expressed as the ratio of daytime CO_2 uptake to daytime transpiration. Canopy (or community) WUE is commonly defined as the ratio of the net CO_2 assimilation of crop canopy to crop canopy transpiratio—that is, the ratio of the canopy CO_2 flux to the H_2O flux for canopy transpiration. Canopy WUE can be expressed continuously and at a daily time-scale, as above, and can also be calculated for specific growth stages. Field WUE can be defined as the 32 ratio of grain yield per unit of water, hence the units would be kg/ha.mm; however the 'plant growth' and 'water' terms need to be explicitly defined. Regional WUE has similar definitions to field WUE except it applies to a larger area. There are three main approaches available to assessing WUE regionally:

1. Remote sensing, which can estimate both evapotranspiration and CO_2 exchange of large areas at specific times of day, can be used to present regional WUE estimates

at specific times. This requires access to much ancillary ground based meteorological data and presents difficulties in extending this from the field sites to regions. There are also difficulties in temporally extending the data from specific times to entire growing seasons;

2. Using remotely sensed based estimates of yield and of λE. However, as has been discussed previously there is difficulty in estimating yield in irrigated environments where LAI is high. Methods exist for providing daily λE maps from specific time-of-day T_s observations, however, reliable estimates of λE are difficult to obtain in highly advective irrigated agricultural environments. Taking isolated daily observations to estimate growing season λE would again require access to ground based meteorological data; and
3. Regional databases of yield, precipitation, irrigation and initial soil water can be developed allowing an 'input-output' definition of regional WUE. 'Input' is the water available over the crop growing season and 'output' is the yield. This approach is well suited to spatial assessment of regional WUE. Coupling this approach with canopy and field-scale process understanding will allow identification of the regional data bases that are required to develop' a more process constrained regional WUE estimation.

For many of the variables that influence WUE, data are usually not regionally available, and hence they cannot be included in the development of regional WUE estimates. Factors varying both spatially and temporally include:

1. crop varieties, which includes plant breeding and genetic modifications;
2. soil conditions, including soil erosion, sodicity, salinisation, and waterlogging;
3. climate change, including precipitation patterns and CO_2 concentration;
4. agricultural practices, including the use of fertilisers, irrigation management, crop rotation, planting density, and the use of mulch to reduce soil evaporation.

These interactions are complex, and largely unknown at the regional scale of southern NSW, making absolute measures of WUE difficult. The 'input-output' GIS definition of regional WUE, however, is most suited to the analysis of relative trends, both spatially and temporally. Such a GIS would rely on access to data recorded by irrigation companies and the RiceGrower's Cooperative, all who are industry participants in the CRC for sustainable rice production.

LIMITATIONS OF REMOTE SENSING IN RICE-BASED AGRICULTURE

Remote sensing is a valuable source of data in rice-based agriculture, especially when regional-scale issues are the concern. However, there are some limitations of remote sensing regarding agricultural applications including:

1. Data availability;

2. Length of recording period;
3. Limited mapping capability;
4. Requirement of expertise and computer facilities; and
5. Cost

Data Availability

Non-availability of remotely sensed data may be due, among other things, to rocket launch, satellite operational problems or political issues. Events such as these will continue to occur and operational systems must pre-determine the influence of any data stream becoming nonavailable. Satellites have different repeat cycles. This means that certain satellites will provide only 2 images per month, for example, whereas others can produce an image every day. This can have considerable impact on agricultural applications since repeat cycle characteristics of satellites are one of the determinants for forecasting yield with the ∫NDVI approach, for example. Also, these high repeat cycle platforms usually have lower spatial resolution, impacting the appropriateness of its data to fine-scale applications. Another major cause of optical data unavailability is cloud coverage. This could be problematic when timing of image acquisition is critical as in crop identification. This may be avoided by using either microwave or airborne remote sensing. However, processing of these data can be a problem due to the scientific expertise needed for analysis of microwave data and the data management associated with large area airborne acquisitions.

Length of Recording Period

The period over which remotely sensed data are available has little impact on agricultural applications. Most agricultural research and management is interested in current or future concerns. However, for the few agricultural projects dealing with historical context, the recording period could be restricting. Remote sensing, unlike meteorological data, has not been recorded for a century. The longest time series currently available of free to ground remotely sensed data covering Australia at monthly time steps is AVHRR at 15 years. LANDSAT data has been recorded since 1972, but can be quite expensive when acquiring a long time series

Limited Mapping Capability

Remote sensing is limited to mapping single crops to slightly higher than 90% accuracy when multi-date, multi-sensor, or GIS data is also 34 used. Remote sensing alone, however, is more limited at discrimination of multiple crops to this level of accuracy. Remote sensing is also often unable to detect direct sources of crop damage. These limitations may decrease in the future as spatial and spectral resolutions, and repeat cycles increase. However, the resolutions needed for this type of detection will probably not be available for some time. Finally, although remote sensing from satellites has high potential for providing spatially variable data needed for precision agriculture, the limitations of fixed spectral bands, too coarse spatial resolutions, inadequate repeat cycles, and long delivery times make it non operational at this time.

Requirement of Expertise and Computer Facilities

Agricultural research with remote sensing requires a moderate level of expertise and computing support. The processing of remotely sensed data requires an investment in training of personnel as well as adequate computers and data storage. Computer *hardware* and *software* are important, but perhaps more important is *mindware* to ensure the correct use of remote sensing to assist in the decision making process.

Cost

Cost of remote sensing projects can be prohibitive, especially when fine detail is needed over large areas. However, this depends on the application and the appropriate remote sensing platform. Much agricultural research has been accomplished with free NOAA AVHRR data. Yet, AVHRR data is not appropriate for many applications. With the launch of the TERRA satellite in 1999, came a new era in remote sensing: no cost moderate to fine resolution data. The spatial resolution of the ASTER and MODIS sensors is appropriate for many different agricultural applications. However, some remote sensing data is currently expensive for large areas of land (e.g., 1,000 km^2), including very high spatial resolution data (e.g., IKONOS or airborne systems), and hyperspectral data. The cost estimates for specified common commercially available remotely sensed data is summarised in Table 3 by each irrigation area in southern NSW. These estimates are intended to give potential users of remotely sensed data a general idea of the costs involved, and by no means are meant as a precise cost. Values in this table are most useful when viewed in the relative context of prices by sensor and by irrigation area.

Table 3: Cost Estimates For Specified Remotely Sensed Data

Irrigation Area	TM	ETM	SPOT PAN	SPOT X	IKONOS[1]
CIA[2]	$750	$600	$1,530	$1,430	$11,172
MIA[3]	$2,375	$2,200	$7,600	$6,800	$63,468
MIL[4]	$3,025	$2,400	$9,500	$8,500	$97,176

[1]*IKONOS* column represents the price for either 1m panchromatic or 4m multispectral data.

[2]*CIA* estimates for TM and ETM, were based on one prices for one ninth scene (up to 3,600 sq km) (map oriented price). SPOT PAN and SPOT X costs are based on one half scene (up to 1,800 sq km) (map oriented price), and IKONOS costs are based on price per area (931 sq km * $12 AUS/sq km = $11,172).

[3]*MIA* cost estimates for TM and ETM are based on one full scene and one ninth scene (full scene is path image price and ninth scene is map oriented price). SPOT PAN and SPOT X costs are based on 4 full scenes (path image price), and IKONOS estimates are based on price per area (5,289 sq km * $12 AUS/sq km = $63,468).

[4] *MIL* estimates for TM and ETM are based on one full scene, one ninth scene and one small scene (path image price for full and map oriented price for ninth and small). SPOT costs are based on 5 full scenes (path image price), and IKONOS costs are based on price per area (8,098 sq km * $12 AUS/sq km = $97,176).

REFERENCES

Anbumozhi, V., Yamaji, E. and Tabuchi, T. (1998). "Rice crop growth and yield as influenced by changes in ponding water depth, water regime and fertigation level". *Agricultural Water Management 37, 241-253.*

Atkinson, P. M. and Lewis, P. (2000). "Geostatistical classification for remote sensing: an introduction". *Computers and Geosciences* 26, 361-371.

Bailey, J. S., Wang, K., Jordan, C. and Higgins, A. (2001). "Use of precision agriculture technology to investigate spatial variability in nitrogen yields in cut grassland". *Chemosphere 42, 131-140.*

Gale, S. J. (1999). "The Snowy Water Inquiry: food, power, politics and the environment". *Australian Geographical Studies 37, 301-313.*

Harrison, B. A. and Jupp, D. L. B. (1989) *Introduction to remotely sensed data,* CSIRO Publications.

Khush, G. S. (1987). "Rice breeding: Past, present and future". *Journal of Genetics* 66, 195-216.

Monteith, J. L. and Unsworth, M. H. (1990) *Principles of Environmental Physics,* Edward Arnold, London.

Seddon, G. S. (1999). Saving the throwaway river. *Australian Geographical Studies* 37, 314-321.

Turner, N. C. (1997). "Further progress in crop water relations". *Advances in Agronomy* 58, 293-338.

Watson, A. S. (1999). "Murray Irrigation Limited". *Australian Geographical Studies* 37, 325-327.

14

GIS Applications in Land Use Analysis

GIS is the systematic introduction of numerous different disciplinary spatial and statistical data, that can be used in inventorying the environment, observation of change and constituent processes and prediction based on current practices and management plans. Remote Sensing helps in acquiring multi spectral spatial and temporal data through space borne remote sensors. Image processing technique helps in analyzing the dynamic changes associated with the earth resources such as land and water using remote sensing data. Thus, spatial and temporal analysis technologies are very useful in generating scientifically based statistical spatial data for understanding the land ecosystem dynamics. Successful utilisation of remotely sensed data for land cover and landuse change detection requires careful selection of appropriate data set. This chapter discusses the land use/land cover analysis and change detection techniques using GRDSS (Geographic Resources Decision Support System) for Kolar district considering temporal multispectral data of the IRS 1C / 1D (Indian Remote Sensing Satellites).

Land-cover refers to the physical characteristics of earth's surface, captured in the distribution of vegetation, water, soil and other physical features of the land, including those created solely by human activities e.g., settlements. Land-use refers to the way in which land has been used by humans and their habitat, usually with accent on the functional role of land for economic activities. It is the intended employment of management strategy placed on the land-cover type by human agents, and/or managers. Landuse/ Land-cover change information has an important role to play at local and regional as well as at macro level planning. The planning and management task is hampered due to insufficient information on rates of land-cover/land-use change. The land-cover changes occur naturally in a progressive and gradual way, however some times it may be rapid and abrupt due to anthropogenic activities. Remote sensing data of better resolution at different time interval help in analysing the rate of changes as well as the causal factors or drivers of changes. Hence it has a significant role in regional planning at different spatial and temporal scales. This along with the spatial and temporal analysis technologies namely Geographic Information System (GIS) and

Global Positioning System (GPS) help in maintaining up-todate land-use dynamics information for a sound planning and a cost-effective decision. Change detection in watersheds helped in enhancing the capacity of local governments to implement sound environmental management. This involved development of spatial and temporal database and analysis techniques. Efficiency of the techniques depends on several factors such as classification schemes, spatial and spectral resolution of remote sensing data, ground reference data and also an effective implementation of the result.

Coastal environment changes were analysed through qualitative evaluation techniques. The techniques included change map derived from vegetation index differencing, Image ratioing, image differencing and image regression. The basic principle of all change detection techniques was that the digital number of one date is different from the digital number of another date. Remotely sensed change detection based on artificial neural networks presents a new technique for multispectral image classification using training algorithm. The trained neural network detected changes on a pixel-by-pixel basis in real time applications. The trained four-layered neural network provided complete categorical information about the nature of changes and detected complete land-cover change "from-to" information, which is desirable in most change detection applications.

Post classification change detection techniques with the comparison of land-cover classifications of different dates have limitations, as it does not allow the detection of subtle changes within land-cover categories. In this regard Open Source GIS such as GRASS (Geographic Resources Analysis Support System) helps in land cover and land use analysis in a cost-effective way. Most of the commands in GRASS are command line arguments and requires a user friendly and costeffective graphical user interface (GUI). GRDSS (Geographic Resources Decision Support System) has been developed in this regard to help the users. It has functionality such as raster, topological vector, image processing, graphics production, etc.

STUDY AREA

Burgeoning population coupled with lack of holistic approaches in planning process has contributed to a major environmental impact in dry arid regions of Karnataka. The Kolar district in Karnataka State, India was chosen for this study is located in the southern plain regions (semi arid agro-climatic zone) extending over an area of 8238.47 sq. km. between 77°21′ to 78°35′ E and 12°46′ to 13°58′ N. (shown in Figure 1) Kolar is divided into 11 taluks for administrative purposes (Bagepalli, Bangarpet, Chikballapur, Chintamani, Gudibanda, Gauribidanur, Kolar, Malur, Mulbagal, Sidlaghatta, and Srinivaspur). The distribution of rainfall is during southwest and northeast monsoon seasons. The average population density of the district is about 2.09 persons/hectare.

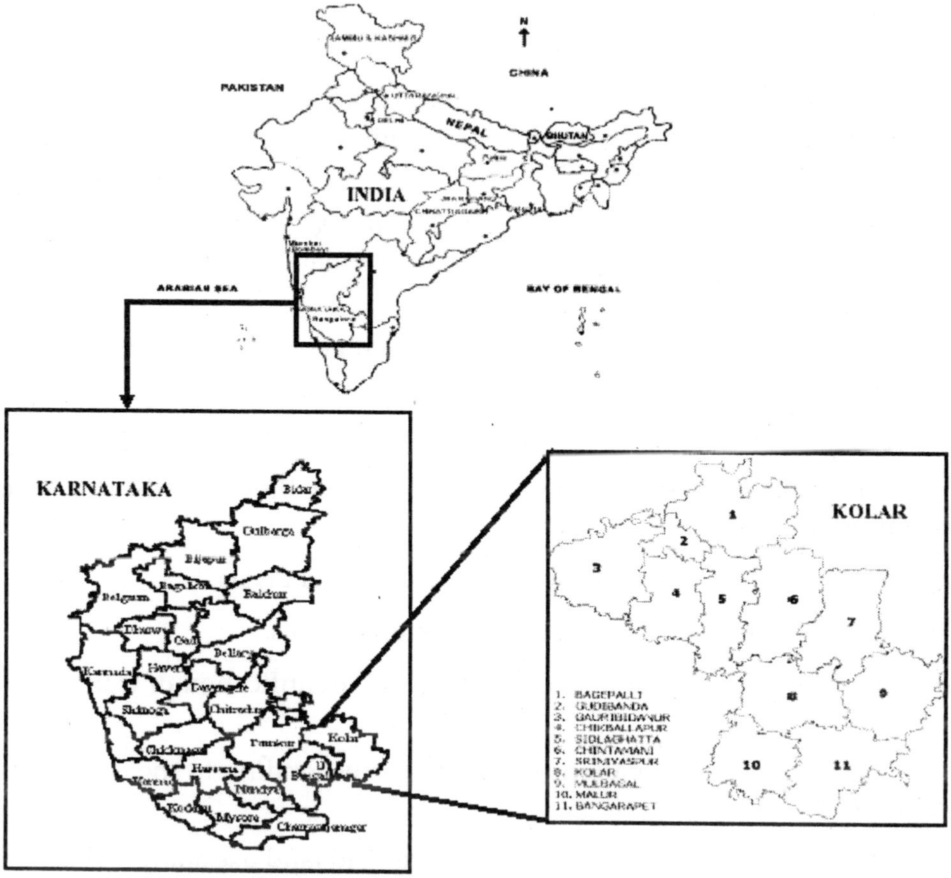

Figure 1: Study area – Kolar district, Karnataka State, India

The Kolar district forms part of northern extremity of the Bangalore plateau and since it lies off the coast, it does not enjoy the full benefit of northeast monsoon and being cut off by the high Western Ghats. The rainfall from the southwest monsoon is also prevented, depriving of both the monsoons and subjected to recurring drought. The rainfall is not only scanty but also erratic in nature. The district is devoid of significant perennial surface water resources. The ground water potential is also assessed to be limited. The terrain has a high runoff due to less vegetation cover contributing to erosion of top productive soil layer leading to poor crop yield. Out of about 280 thousand hectares of land under cultivation, 35% is under well and tank irrigation. The main sources of primary data were from field (using GPS), the Survey of India (SOI) toposheets of 1:50,000, 1:250,000 scale and multispectral sensors (MSS) data of the IRS (Indian Remote Sensing satellites)-1C and IRS-1D. LISS-III MSS data scenes corresponding to the district for path-rows (100,63) (100,64) and (101,64) was procured from the National Remote Sensing Agency, Hyderabad, India. The secondary data was collected from the government agencies (Directorate of census operations, Agriculture department, Forest department and Horticulture department).

METHODOLOGY

The methodology of the study involved:

1. Creation of base layers like district boundary, district with taluk and village boundaries, road network, drainage network, contours, mapping of waterbodies, etc. from the SOI toposheets of scale 1:250000 and 1:50000.
2. Extraction of bands (LISS3 with resolution 23.5 m and PAN with resolution 5.8 m of 1998 and 2002) from the data (in BIL and BSQ format) respectively procured from NRSA.
3. Identification of ground control points (GCP's) and geo-correction of bands through resampling.
4. Cropping and mosaicing of data corresponding to the study area.
5. Fusion of LISS3 and PAN data using RGB (Red, Green, Blue) to HIS (Hue, Intensity, Saturation) and HIS to RGB conversion technique.
6. Histogram generation, Bi-spectral plots, Regression analysis.
7. Computation and analysis of various vegetation indices.
8. Generation of FCC (False Colour Composite) and identification of training sites on FCC.
9. Collection of attribute information from field corresponding to the chosen training sites using GPS.
10. Classification of remote sensing data: Land cover and land use analyses (both district wise and taluk wise).
11. Change detection analysis using different techniques (Image differencing, Image ratioing, etc.).
12. Detection, visualisation and assessment of change analysis.
13. Statistical analysis and report generation.

RESULTS AND DISCUSSION

Land cover analysis was done by computing Normalised Difference Vegetation Index (NDVI) which shows 46.03 % area under vegetation and 53.98 % area under non-vegetation. Vegetation index differencing technique was used to analyze the amount of change in vegetation (green) versus non-vegetation (non-green) with the two temporal data. NDVI is based on the principle of spectral difference based on strong vegetation absorbance in the red and strong reflectance in the near-infrared part of the spectrum.

$$D_{NDVI} = (IR - R)/(IR + R)\ t_2 - (IR - R)/(IR + R)\ t_1 \qquad (1)$$

t_1 and t_2 in the equation denote the two different dates, where t_1 is for the year 1998 and t_2 for 2002. The result shows a 16.46 % difference in the vegetation area between the two dates. Land use analysis was done by both Supervised classification (accuracy 94.67 %) and unsupervised classification approach (accuracy 78.08 %) using Gaussian Maximum Likelihood Classifier (GMLC) to classify the data in to five categories (agriculture, built-up, forest, plantation and waste land).

The Land use analyses as given in table 1, indicates increase of non-vegetation area from 451752 ha. (54.84% in 1998) to 495238 ha (60.17% in 2002). The results also show decrement in forest area and increment in builtup (18.79 %), plantation (12.53 %) and waste land (41.38 %) in 2002 against that in 1998 (builtup-15.96%, plantation-8.53% and waste land-38.88%). Further, taluk wise land use data was extracted by overlaying taluk boundaries and results are tabulated in Table 2.

Table 1: Land use details of Kolar district Year 1998 Year 2002

Categories	1998 Area (in ha)	1998 Area (%)	2002 Area (in ha)	2002 Area (%)
Agriculture	233519	28.34	165711.42	20.13
Builtup	131468	15.96	154668.68	18.79
Forest	68300	8.29	58979.35	7.17
Plantation	70276	8.53	103110.13	12.53
Waste land	320284	38.88	340570.16	41.38

Table 2: Taluk wise land use in percentage area (1998 and 2002)

Taluk	Agriculture (%)		Built up (%)		Forest (%)		Plantation (%)		Waste land (%)	
	1998	2002	1998	2002	1998	2002	1998	2002	1998	2002
Bagepalli	15.75	12.69	22.46	44.65	09.26	03.28	03.65	07.51	48.88	31.86
Bangarpet	27.43	14.15	15.83	09.65	15.95	12.59	13.97	13.32	26.82	50.28
Chikballapur	30.61	30.28	10.56	13.59	18.30	13.35	08.16	15.18	32.37	27.60
Chintamani	29.94	20.07	13.59	20.11	01.95	01.61	05.52	08.52	49.00	49.69
Gauribidanur	22.75	17.24	22.11	23.97	06.50	04.12	02.61	11.57	46.03	43.10
Gudibanda	15.58	22.71	11.04	19.69	04.47	04.67	02.55	09.42	66.36	43.52
Kolar	33.47	21.81	13.09	12.93	05.70	08.62	07.67	14.25	40.07	42.40
Malur	40.95	22.56	08.52	12.84	03.03	09.05	19.62	17.12	27.88	38.42
Mulbagal	22.85	19.26	21.13	12.72	06.25	01.98	09.35	06.58	40.42	59.46
Sidlaghatta	32.47	24.72	13.95	24.76	03.27	07.61	10.75	15.92	39.56	26.98
Srinivaspur	36.52	22.93	15.34	08.04	13.65	12.67	09.34	19.10	25.15	37.25
District	28.35	20.13	15.96	18.79	08.29	07.17	08.53	12.53	38.87	41.38

LISS3 multispectral (MSS) data of the IRS 1C and 1D of resolution-23.5 meters (both 1998 and 2002) were merged with the PAN data of IRS 1C resolution-5.8 meters using the HIS fusion technique for better spatial and spectral resolutions. HIS fusion converts a color image from the RGB (Red, Green, Blue) space into HIS (Hue, Intensity, Saturation) colour space. The intensity (I) component resembles a panchromatic image, and hence is replaced by a panchromatic image of better spatial resolution. A reverse HIS transformation of the panchromatic together with the hue (H) and saturation (S) bands, result in the fused image. Supervised classifications were performed for selected taluks with ground truth data. The comparative results of the taluks where subtle change detection could be observed in 2002 are as listed in Table 3 and the corresponding taluk wise area in percentage are as listed in Table 4.

Table 3: Talukwise land use area in hectares (ha) of the year 2002

Taluk	Agriculture	Built up	Forest	Plantation	Waste land
Chikballapur	19220.54	9293.13	7143.66	9099.19	19064.50
Chintamani	19958.61	19957.48	1488.25	7358.55	40140.64
Gauribidanur	15612.86	19447.85	3310.56	10929.94	39557.35
Gudibanda	5080.74	4662.85	846.32	2738.80	9398.59
Mulbagal	13251.53	10034.88	2578.21	4940.57	51168.42
Sidlaghatta	15872.94	13614.46	5145.42	12425.18	19999.06
Srinivaspur	20189.23	7650.96	11006.07	15490.01	31942.53

Table 4: Taluk wise land use in percentage area (1998 and 2002)

Taluk	Agriculture (%)		Built up (%)		Forest (%)		Plantation (%)		Waste land (%)	
	1998	2002	1998	2002	1998	2002	1998	2002	1998	2002
Chikballapur	32.08	30.12	08.57	14.56	17.55	11.19	10.97	14.26	30.82	29.87
Chintamani	23.45	22.45	12.90	22.45	04.22	01.67	08.13	8.28	51.00	45.15
Gauribidanur	25.46	17.57	21.43	21.89	07.98	03.73	02.77	12.30	42.36	44.52
Gudibanda	16.71	22.36	12.63	20.52	05.25	03.72	03.29	12.05	62.12	41.35
Mulbagal	23.23	16.17	20.68	12.24	06.59	03.15	09.37	06.03	40.13	62.42
Sidlaghatta	30.94	23.67	15.18	20.30	03.12	07.67	09.94	18.53	40.83	29.82
Srinivaspur	33.39	23.40	17.97	08.87	10.29	12.76	09.70	17.95	28.64	37.02

Comparison of the temporal data shows that builtup has considerably increased in Chikballapur (14.56 %) showing urban sprawl in and around the center of the town at the road junction and the forest area has decreased by 6.36%.

Change Detection Techniques

Different change detection techniques such as image differencing, image ratioing, vegetation index differencing and Image regression were attempted to assess the amount of change in the study area.

Image Differencing

Georeferenced images of two different time periods t_1 and t_2 were subtracted on a band by band and pixel by pixel basis to produce an image which represents the change between the two time periods.

$$D_{x\,ij}^{k} = X^{k}_{ij}(t_2) - X^{k}_{ij}(t_1) + C \tag{2}$$

where, X^k_{ij} = pixel value for band k and i and j are line and pixel numbers in the image, t_1 = first date and t_2 = second date and C = a constant to produce positive digital numbers.

This technique takes into account the difference of radiance values of pixels between two different dates. Differences in atmospheric condition, differences in sensor calibration, moisture condition, illumination condition also affect the radiance of the pixels. Therefore this technique is better suited to cases as changes in radiance in the object scene is larger compared to changes due to other factors. Frequency analysis of the image show that the pixels with the radiance are found in the tails of the distribution while non-radiance change pixels tend to be grouped around the mean. Figure 2 shows the histogram obtained for the band 4 (near infrared) from image differencing.

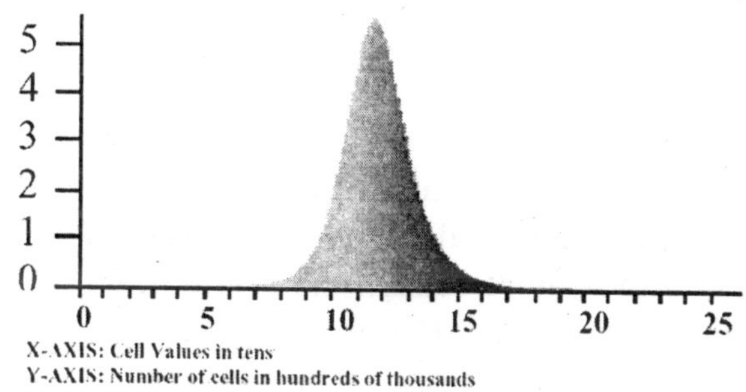

Figure 2: Histogram of the near-infrared band obtained from image differencing

The histogram of the difference image with an ample amount of pixels in the tails clearly indicates changes. However the actual change was unpredictable due to lack of detailed ground truth data pertaining to different categories.

Image Ratioing

Geocorrected images (G, R and NIR bands) of different dates were ratioed pixel by pixel (band by band) basis.

$$Rx^{k}_{ij} = X^{k}_{ij}(t_2)/X^{k}_{ij}(t_1) \tag{3}$$

Where, $X^k_{ij}(t_2)$ is the pixel value of band k for pixel x at row i and column j at time t_2. If the intensity of reflected energy is nearly the same in each image then $Rx^k_{ij} = 1$ indicating no change.

The ratio value greater than 1 or less than 1 represents a change depending upon the nature of changes occurred between the two dates. Figure 3 shows the histogram obtained for the near infrared band after performing the image ratioing. The histogram generated for the different bands showed that a significant part of the image has no change as the number of pixels falling to the category '1' was dominating (with a high peak in the histogram) compared to pixels that had values greater than or less than '1'.

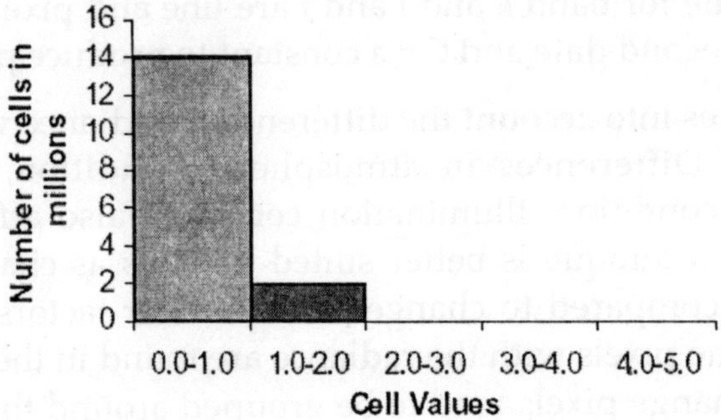

Figure 3: Histogram of the NIR band after temporal date images ratioing

Image Regression

It is assumed that pixels from time t_1 are in a linear function of the time t_2 pixels. Using linear regression $X^k_{ij}(t_1)$ was regressed against $X^k_{ij}(t_2)$. It accounts for the differences in the mean and variance between digital number of pixels of different dates in order to reduce the differences in atmosphere condition or Sun angle. The difference image DX^k_{ij} is given with the predicted value $X^{\wedge k}_{ij}(t_2)$, is given by

$$DX^k_{ij} = X^{\wedge k}_{ij}(t_2) - X^k_{ij}(t_1) \quad (4)$$

Regression analysis was performed for each bandand the results are listed in Table 5.

Table 5: Image regression results

X (1998) (Independent variable)	Y (2002) (dependent variable)	S	I	r	R	P (Significant value)	DN (Digital number)	D (difference)
Band 2	Band 2	–0.43	149.62	–0.636	0.404	< 0.011	128	–33.42
Band 3	Band 3	–0.55	151.80	–0.694	0.482	< 0.004	117	–29.55
Band 4	Band 4	–0.67	172.17	–0.546	0.298	< 0.035	111	–13.20

S = slope, *I* = intercept obtained from linear regression, r = the correlation coefficient, R = coefficient of determination, *P* = significant value or significance level, *DN* = digital number of the pixel that was chosen from the set of numbers in the bands and *D* = the predicted value that would be obtained from equation 4.

In this method the mean and variance of the pixels takes care of environmental interferences like adverse effects from atmospheric condition and sun angle by distributing these variations to all the pixels. Thus the differences obtained in this analysis is minimal when compared to other methods.

REFERENCES

Dai, X.L., and Khorram, S., Remotely Sensed Change detection based on Artificial Neural Networks. *Photogrammetric Engineering & Remote Sensing,* 65(10), pages 1187-1194, 1999.

Macleod, R.D., and Congalton, R.G., A quantitative comparison of change detection algorithms for monitoring eelgrass from remotely sensed data. Photogrammetric Engineering & Remote Sensing, 64, pages 207-216, 1998.

Mitra, D., Change detection using remote sensing data in a coastal environment. *Trendsin Geoinformatics Technology and Applications, pages 107-112, 1999.*

Prenzel, B., and Treitz, P., Remote sensing change detection for a watershed in north Sulawesi, Indonesia. *Progress in Planning,* 61, pages 349-363, 2004.

Xavier, B., and Gerard, S., LUCC data requirements workshop. *Survey of needs, gapsand priorities on data for land-use/land-cover change research, Barcelona, Spain, LUCC* Report Series No. 3, 1997.

15

Applications of Intelligent Archive Systems

NASA expects to accumulate very large volumes of earth and space science data in the next decades as a consequence of advances in observational instrumentation capabilities. More powerful space-based and terrestrial sensors (ranging large to nano-scale) including new strategies for deploying distributed and networked sensors will be responsible for a host of observations rendered as digital data. Managing these data while simultaneously extracting the most societal value from them pose significant challenges for designers of future archive systems. Studies that look deeply into the next generation of data archiving systems are yielding new concepts to address these challenges. A study team funded under the NASA Ames CITC Intelligent Systems programme has been developing the concept of an Intelligent Archive (IA). The IA team studies the role of an IA within the context of an end-to-end scientific enterprise. This context embraces a chain of systems extending from those involved in data acquisition to those that derive information and knowledge from those observations and measurements for use by various societal enterprises. Implications of terabyte and peta byte-era data volumes on systems and infrastructures that efficiently support science-based applications stimulated research into new concepts for intelligent archives in the years 2012-2020. Beyond the obvious dependencies on improvements in hardware technologies to meet requirements for storing and supporting efficient access to future distributed data volumes, advances are needed in concepts and tools to enable intelligent data management and utilisation. While the growth rate and quantity of data increases rapidly, new challenges are posed to basic data management and utilisation. Key challenges that characterise future data volumes and use include:

1. Data acquisition and accumulation rates tend to outpace the ability to access and analyze them
2. The variety of data implies a heterogeneous and distributed set of data providers that serve a diverse, distributed community of users
3. Unassisted human-based manipulation of vast quantities of archived data for discovery purposes is difficult and potentially costly

4. If NASA is to migrate its technologies to operational agencies' decision support systems, it is necessary to demonstrate the feasibility of near-real-time utilisation of vast quantities of data and the derived information and knowledge
5. The types of data access and usage in future years are difficult to anticipate and will vary depending on the particular research or application environment, its supporting data sources, and its heritage system infrastructure

The IA team has developed a conceptual architecture with new intelligent system-based capabilities for future archives to address these challenges. Part of the process for developing IA concepts included the use of scenarios describing data usage in applications combined with projections of advances in relevant technologies. In this manner an abstracted architecture could be defined without regard to physical implementation. Architecture for future archives can therefore be considered from the point of view of the functions that need to exist in support of various usage scenarios and overall requirements. This points out that the functions of an intelligent archive are more stable than the physical architectures and technologies used to implement them. By 'discovering' and abstracting required components and processes from scenarios into functional elements. An outgrowth of this approach is a definition of characteristics that describe future intelligent archives. An IA is different from contemporary archives because the scope of meaning for the term 'archive' is extended from a simple repository of data to one that supports and facilitates derivation of information and knowledge. It supports greater data utilisation integral to knowledge building systems. For example, items managed by an IA include:

1. Data, information, and knowledge
2. Software needed to manage holdings
3. Interfaces to algorithms and physical resources to support acquisition of data and their transformation into information and knowledge, storing the protocols to interact with other facilities

With the application of intelligent algorithms and intelligent system components, an IA has greater ability to operate more autonomously than conventional archives. This characteristic reflects higher degree of automation and self-managed operations. Accordingly, an IA is envisioned to provide highly responsive and accurate services to users (e.g. as an intelligent assistant) with less operator intervention. By embodying new capabilities like these an IA can be distinguished from archives of today with regard to its ability for:

1. Storing and managing full representations of data, information, and knowledge
2. Building intelligence about transformations on data, information, knowledge, and accompanying services involved in a scientific enterprise
3. Performing self-analysis to enrich metadata that adds value to the archive's holdings

4. Performing change detection to develop trending information
5. Interacting as a cooperative node in a "web" of other systems to perform knowledge building (where knowledge building involves the transformations from data to information to knowledge) instead of just data pipelining
6. Being aware of other nodes in the knowledge building system (participating in open systems interfaces and protocols for virtualisation, and collaborative interoperability)

While internal capabilities of the IA could be characterised and described like those above recognised that a broader contextual perspective was necessary to further evolve the concept. For instance, the role of an IA in an environment of data providers and data and information consumers remained unclear. Questions about how an IA interacts as an interoperable and cooperative component in an overall distributed system environment proved initially ambiguous. We elected to develop use case scenarios to study how different application users could benefit from IA capabilities end-to-end.

USE CASE SCENARIOS FOR IA CONCEPTS

Use case scenarios help illustrate, exercise, and clarify IA concepts. Scenario-based approaches to developing intelligent archive concepts also help derive clearer understanding of end-to-end interrelationships among data and information, consumers, data providers, value-added information services, data archives, and data acquisition missions that underlie an IA. Furthermore, use-scenarios provide better insights into limitations of existing solutions, and the nature of challenges, issues, and probable opportunities for innovation. Scenarios of conceptualised ways to use intelligent archive resources help uncover a range of requirements for services and capabilities that can be mapped to existing and future technology applications. Consequently, forward looking, tangible and imaginative scenarios lead to a flexible, more comprehensive architecture for future intelligent system applications. By starting with an existing realistic set of requirements explored ideas from which to extrapolate the possible and then develop more speculative and visionary scenarios. Each was studied with regard to advanced data/information requirements and the challenge of assimilating progressively massive amounts of available data. Studying a conceptual application domain contributes to an appreciation of both context-specific requirements and those for common or generalised features. First considered the current domain for data archive systems. With a basic model of data archive systems operating in a scientific knowledge building domain where data acquisition, preservation, access, distribution, and utilisation are involved, related nearterm, next-generation, and future requirements to use case scenarios. The generalised model of common features that emerged from this study consists of four enterprise elements or tiers:

1. *Sources:* includes missions, instruments, and platforms for geo-spatial observations acquired by data receiving stations.

2. *Providers:* includes primary data processing, secondary processing and product production, services, archiving, and access/distribution channels.
3. *Value Chains:* includes secondary and other data product provider service groups as part of end-to-end value-added services.
4. *Consumers:* includes data and information users possessing challenging requirements for geo-spatial data products/services.

Each tier represents a high level stakeholder in the enterprise that has its own evolving requirements for information technology capacities (e.g., throughput, storage volumes, and communication bandwidth), capabilities, inputs, outputs, and interfaces. Interdependencies and various patterns of relationships (e.g. functional, transactional, and operational) exist within and among the enterprise tiers. Evolving requirements among the enterprise elements affect the nature of these interrelationships and dependencies. By considering the implications requirements for future archives have on supporting technologies and, conversely, how technological advances might be exploited for interoperable elements of enterprise systems.

PRECISION AGRICULTURE USE CASE STUDY

Precision agriculture is concerned with techniques and practices that maximise crop yield with minimum cost through improved use of data, information, and knowledge technologies. Explore precision agriculture from the perspective of the consumer or "enduser" (e.g. fourth tier stakeholder). Special attention is given to data and information services useful in precision agriculture involving an individual farm. Additional stakeholder viewpoints on requirements include agribusiness, various government agencies, and research communities. By beginning with a single farm it is possible to study scaled up scenarios involving a grid of connected farms aggregated into regional, national or even global levels of cooperation. We start with a use case scenario based on the scope and parameters of a minimal agricultural production unit with regard to planning, cultivating, and harvesting crops.

Use Scenario: Grower And Farm

The farm is characterised as a relatively small spatial area (considered in acres) for agricultural products suited to regional ecological, weather, and growing constraints. Examples of farms vary regionally such as a Midwestern 360 acre farm for corn and soybeans; a South Dakota 2000+ acre wheat field; an average California Central Valley 800 acre vegetable and fruits farm; and a Montana 10000+ acre cattle ranch. Geo-spatial information relevant to these farms puts a premium on high-resolution data both spatially and temporally. We will use the scenario to generate challenging requirements with which to exercise the conceptual IA architecture and better understand the dynamic interactions among its components and functions.

Planning, Cultivating, Harvesting

Quality data, information, and knowledge are highly important for conducting precision farming activities like crop planning, cultivation, and harvesting. Information-intensive support services required for each activity include monitoring current conditions, collecting histories and time series studies, running trends/risks analysis, making predictions, forecasts, "what-if" investigations, and harvest outcome comparisons. Detailed information about land, weather, water, agriculture markets, prior yields, agricultural-chemical options, seeds, etc. is required for planning crop selection and planting. High-resolution information is required to monitor crops, assess risks, and make decisions about appropriate interventions to maintain crop health. Similarly, to maximise yields, decisions about timing harvests require information about current and future conditions. Remotely sensed information about farm assets including information collected from the farm about outcomes of plans, cultivation techniques, and harvests, is integrated within a farm's database for perpetual use.

Vision Concept: Digital Farm Interface

Humans perpetually invent ways to reduce work, or at least to make it more manageable. While mechanical machines alleviate physical work, digital machines address information-age burdens. The future of farming exploits advantages from both types of machines. Farming represents a dynamic interplay among machines, information, and human knowledge-based techniques employed in the critical pursuit of food production. Precision farming research is one such frontier where applying information technologies promises to improve operational efficiency while simultaneously maximising crop yields. The concept of a digital farm merges applications of information technologies, digital information, GIS, GPS, archives, value added-providers, sensors, and human-machine interfaces (see Figure 1).

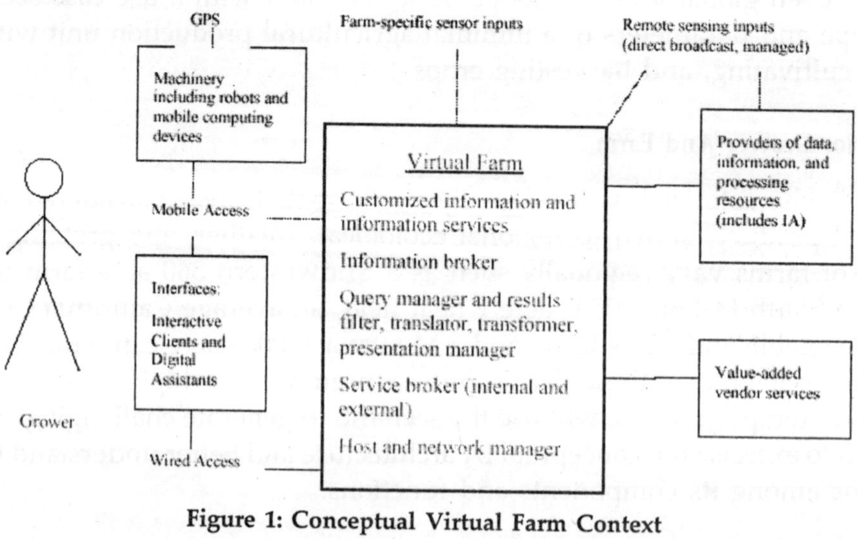

Figure 1: Conceptual Virtual Farm Context

Quality high-resolution geo-spatial data are integral to precision agriculture and the decision-making processes practiced within it. The following speculative scenario describes how future growers might access and use specialised information selected from vast data stores that are constantly updated with hourly observations. The grower's information resources represent the consequence of interoperating services, value-added processes, automation, and filtering of data of specific relevancy to the farmer.

Digital Farm Vision Scenario

When a South Dakota wheat grower rises early each morning, part of his or her morning routine involves perusing world news, local news, and a "news" review of the entire wheat farm. Summary and detailed information about every aspect of the farm, including past, present, and future, is made easily available through interactive high-resolution panels. A digital assistant or "agent" works on behalf of the grower to collect and provide valued information services. In the background mode this artificial assistant actively monitors sources of data and information for updates to harvest and constantly update in the digital farm database. An interactive mode is also available for the grower to speedily access ad hoc information. The digital assistant is accessible from any interface (e.g. workstations and mobile devices) within the house, farm buildings, vehicles, or even the combine. Interaction with the digital assistant can be conducted by natural language via voice or touch screens and keyboards.

"Intelligent" digital assistant software can interpret, broker, and fulfill requests for information dynamically and autonomously. This feature of a digital farm skillfully performs its role by brokering and invoking services from a digital virtual wheat farm database that contains encyclopedic farm-relevant information ontologically, spatially, and temporally organised. The digital farm automatically ensures that its information resources about soil, crop, weather, and moisture conditions are constantly updated. To do so digital farm software has interfaces with external data and information sources for input as well as with farm-specific sensor inputs (see figure 1 Virtual Farm links). These functional interfaces are crucial to pooling farm-relevant data from multiple data sources such as primary archives and agricultural "value-added" services. The digital assistant can also produce different views of this information by summoning an array of digital functional services. These virtual farm services produce virtual 4-D model (i.e. topographic with temporal perspectives) of the entire farm that the grower can inspect from his or her office or combine cab monitor for example. The virtual farm serves as an interactive reference of farm-specific assets integrated with historical, current, and modeling information. Views of the farm can be summoned to within a square meter with variable time sequences. Types of information range from historical to actual current conditions to what-if scenarios cast into future periods of interest. Because the grower's digital farm can "learn" from his or her queries and interests, the content and services it provides adapt with change and specificity over time.

Some of the farm machinery also interacts with the digital farm information services (see figure 1). Autonomous and semi-autonomous machines that plant, cultivate, and harvest crops are precisely controlled with a combination of GPS, distributed functions, and data from the digital farm. Optimal applications of seeds, fertilisers, and chemicals can be controlled and recorded via wireless digital farm services. Similarly, data taken from the field during cultivation and harvesting can be relayed to the digital farm as input for archiving and further use. Overall the digital farm concept represents a sophisticated intelligent data and information pool. It serves as an interface between specific interests about the farm and vast external data sources. It also serves to integrate and administer different kinds of extracted (i.e. mined) and filtered information. Digital farms also provide the basis for intelligent interfaces for machine-to machine (e.g., smart farm machinery, digital assistants) and human-machine interactions.

Digital Farm: Consequences

We speculate that within fifteen years, digital information services and systems will be integrated into many automated agricultural functions. Farm functions requiring improved accuracy, risk mitigation, production optimisation, safety, environmental, and regulatory compliance will be managed better with higher levels of integration with information services. The agricultural production unit could also be integrated into a collaborative and adaptive system that efficiently shares information for the benefit of all agricultural enterprise stakeholders. In the next decades both *in situ* and remote sensing observations/measurements should be smoothly interfaced with data assimilation services that support a variety of private, public, and cooperative concerns. Demonstrated improvements in production and quality of agricultural products continue to be realised, thus reinforcing further evolution of the agri-information and agri-business enterprise.

Another consequence of digital farms includes how they usher in an era characterised as highly integrated agriculture—digital farms containing local precision agricultural data and histories could be subsumed into a connected grid of many farms aggregated into county, state, regional, national, and perhaps even global farming sectors. An agricultural grid of services and interchangeable information could support many levels of functional cooperation involving members with individual operational parameters to those of large collective organisations. Furthermore, it is conceivable that such a grid of agricultural stakeholders could itself be interfaced with economic grids that link markets to enterprises in new ways. It may be common practice in this future time for interconnected agri-economies, businesses, and world markets to transact business based on clearer pictures of actual crop production status summarised at various levels of aggregation down to the individual production unit.

SCOPE OF DIGITAL FARM REQUIREMENTS

Data from which to produce information used in the digital farm scenario are scoped to the physical parameters of an agricultural production unit. This means discrete agricultural production units are defined as relatively small geographic areas with specific boundaries nested into micro-ecological locales, regional habitats, and environments. Remotely sensed information must be relevant to the agricultural enterprise practiced by the grower and related to the agricultural assets of the farm. Precision agriculture relies on accurate, quality information keyed to geo-spatial resolutions of a meter or finer (to help differentiate among plants and distinguish plant conditions within the farm). Temporal resolutions often need to be hourly especially for weather information that is tailored for the locale and region of interest. Precision weather forecast requirements range between three hours ahead to months and perhaps even years. Examples of additional requirements for the content of a digital farm include:

1. Land profiles that include soils, moisture, elevations, drainage patterns, and water sources.
2. Digital Ortho-photo Quadrangles (DOQ)—digital images that coincide or integrate USGS topography maps—1 meter resolutions
3. Digital Elevation Models (DEM)
4. Geo-rectified spatial data
5. Ecological zone profiles including habitat, botanical profile, zoological profile, and supportable species including plants, insects, bacteria, viruses, animals, etc.
6. Inventory of actual biological diversity
7. Maps of contiguous environmental components and configurations
8. Local calibrations including ground truth for valued and relevant agricultural assets with regular adjustments
9. Local calibrations including ground truth for weeds, pests, plant-stress signatures, etc.

These requirements suggest that the content of a digital farm is largely composed of georegistered land and weather data scaled to the physical dimensions of the farm and its immediate environment. Various services and functional capabilities are therefore required to acquire and transform data from various resources into information relevant to a particular farm. Part of this 8 relevancy includes information that supports primary operations of the precision farming enterprise.

Required Services and Capabilities

For a virtual farm to be an effective service to farmers it must possess a variety of functional capabilities that support precision agricultural operations. Key among those is the overall capability to broker modes of access to distributed sources of data and information. A virtual farm would be capable of invoking services sufficient to acquire, filter, and transform data into farmrelevant information. These include the capability to:

1. Extract relevant information from data scaled to the interests of the specific farm
2. Extract farm-specific information from models for forecasts and "whatifs" projections
3. Provide rapid turnaround for requested data/information ranging from complex to simple
4. Broker information service requests with multiple service providers
5. Provide easy to understand graphical presentations of information
6. Communicate via wireless and wired connections

A variety of specialised services are also encompassed and provided by a digital farm. They play a role in supporting farm management activities with dynamic and interactive access to various data/information resources that can be chained into delivering customised information products. Examples of useful services identified in our study of the use case are listed below:

1. Planting conditions indicators—predictive, actual, recommendations, interpretations
2. Crop monitoring—daily observations that can be stored, interrogated, examined at various levels of detail, summarised, and shared
3. Crop growth measurements—predictive, actual, interpretations
4. Crop maturity—daily projections, actual
5. Plant health indicators (stress, pests, infections)—actual, optimal, projected
6. Identification and tracking of weeds and other infestations—type, growth, condition, spread, actual, projected
7. Soil moisture indicators—historical, current and long-term, optimal vs. actual.
8. Chemical dispersion impact—actual distribution correlated with other components 9Weather conditions—hourly monitoring of wind, temperature, precipitation, cloud cover
9. Advance forecasts—daily, weekly, monthly, seasonal, annual, multi-year climate change
10. Environmental alerts –concerns such as storm, hail, drought, fire, flood, freezes, snow

11. Sunshine vs. cloudy predictors—correlation of projected weather with crop type options
12. Microclimate zones and elevations—comprehensive survey of farmland environment
13. Soil types and depths—mapping, survey, and correlation with ground cover and optimal crop types
14. Plant census—distributions by type and location; current and historical
15. Supportable biology—potential vs. actual; daily, seasonal, trends
16. Stress signatures—causes and signs
17. Interventions—chemical applications; actual, planned, historical, outcome analysis

From this list ascertain that precision agriculture requires specific kinds of information for a variety of topics required throughout an operational year of a farm. Furthermore, such information in many cases needs to be extracted, derived, and processed to aid decision support. Timeliness, quality, and resolution of utilised information factors significantly into the quality of decision making expected when growers address issues confronting each aspects of sustaining farm productivity.

Implications of Overall Requirements on Collaborative Precision Agriculture Stakeholders

Growers equipped with a comprehensive agricultural GIS resource (i.e. digital farm) custom oriented to the parameters, dimensions, and location of the farm, will depend on internal (farm sensors) and external sources for data and information services (e.g. NASA, non-NASA, and value-added providers). Principal among these sources would be intelligent archives and value-added providers of agriculture information services. Resource service providers of data and information would be required to make their service available twenty four hours a day by subscription and via dynamic requests (i.e. initiated by the grower, agents, and digital assistant).

Data and information is the common denominator for stakeholders in the precision agriculture enterprise. The flow of Information will be characterised more like a federation of sources and services that represent innovative, basic, value-added, and competitive services within and from outside the agricultural enterprise. For example, intelligent data archives of government maintained remote sensing data could supply data directly to a virtual farm or to value-added information providers. Distributed and collaborative intelligent archives could exist among government, commercial, institutional (university), and private organisations that support virtual farms with information services. Increasingly, stakeholders will recognise the value of remote sensing and in-situ data for precision agriculture. This will encourage political and economic forces to define the evolution of stakeholder services and interoperations among them.

INTELLIGENT ARCHIVE ROLE

An IA is intended to address relatively large-scale operational and functional data management and utilisation requirements (e.g. primary remote sensing data archives, mission or discipline specific archives, or custom commercial archives). Analyses of requirements from the precision agriculture use case context suggest a virtual farm's value stems from tailoring and customising information germane to a farm. Information used by a farm will depend on applications of intelligent data understanding to extract, transform, and present the information it needs. A significant role for an IA is in distilling specific farm-relevant information from an avalanche of data. These transformations may involve complex process chains of data reduction and information/knowledge extraction services. Functional components for an IA to facilitate the distillation of information for a farm include:

1. Instrument/sensors for resolving farm-scale features and assets
2. Hyperspectral data mining and extraction techniques
3. Machine learning for automating the evolution of system intelligence
4. Modeling and predictions
5. Algorithms for data products
6. Visualisation of preprocessed and dynamically processed data

These component technologies must be designed and implemented to interoperate throughout end-to-end processes for data production, quality assurance, cataloging, metadata extraction, data content characterisation, data product archiving, and custom utilisation tools. Automated data understanding is crucial for distillation, characterisation, detection, and extraction of such things as objects, phenomena, features, conditions, and other knowledge-based subjects. Data understanding within an IA can be illustrated another way with regard to alerts. Intelligent data understanding provides the means to automatically detect, recognise, and assess conditions or phenomena of concern, automatically formulate an alert message (with visualisation as appropriate), automatically determine who should receive it, and send it via the most expedient channel(s). An IA will also provide an interface through which it interacts and collaborates with other intelligent systems. One example involves system intelligence to accept and process semantic queries or broker requests to appropriate internal/external services. A chain of semantic filters and information retrievers comprise an IA functional layer for transacting requests. This layer may also filter relevant information on behalf of people (e.g. grower and crew) and present unambiguous, relevant information through interactive interfaces tailored to the concerns of the 11 consumer. Additional implications of virtual farm requirements for an IA to address, considering an end-to-end perspective, include:

1. Visualisation, mining, analysis, and discovery tools support

2. Digital virtual interfaces
3. Digital assistants
4. Query/result agents
5. Interfaces to the Semantic Web
6. Interfaces to a knowledge grid
7. Distributed knowledge, information and data
8. Archive-to-archive interoperability
9. Distributed computation
10. Distributed services
11. Data intensive computing
12. Remote analysis services
13. Massive data volumes and complexities in data
14. Responsiveness versus costs to extract useful knowledge and information

VIRTUAL FARM QUANTITATIVE DATA REQUIREMENTS ASSESSMENT

Farms vary by size and production orientation (e.g., wheat vs. cattle, fruit vs. soybeans for example). Data requirements are estimated for a 1000 acre area farm with which variable size farms can then be calculated (e.g., actual acres/1000 x data estimates in bytes). Estimation of data requirements is difficult because they depend on various factors such as the number of sensors, data sources used, parameters required, and frequency of sampling. Some assumptions were required. Assessments of data for virtual farms are derived from several categories of data and information resources influenced by frequency of availability and need. Some initial working assumptions are:

Farm size: 1000 acres {Acre = 4047 m^2 (63.8 × 63.8m) * 1000
acre farm = 4047000 m^2 area = ~4km^2 area (or 2 km × 2 km farm)}

Remote sensing spatial resolution requirements: 1m, 10m, wider-range

Temporal resolutions: hourly, daily, history, 1-3 hour and longer-range measurements and forecasts

In situ sensors: mobile mounted, static, configurable (e.g., variable densities)

Parameters: soil conditions, plant conditions, weather conditions, etc.

Information Contexts: specific farm, local context for farm, regional contexts (includes interdependencies among micro-zones, eco-zones, growing zones, weather patterns, etc.)

Categories of Virtual Farm Input Resources

The envisioned virtual farm utilises a variety of data and information resources. These can include whole data inputs (whole scene or granule), subset data, preprocessed data, modeling results, subsets of larger modeling results, visualisation products (such as animations), or preprocessed data to support interactive visualisations and merging with other data types. Correspondingly, quantification of data requirements will be variable due to differences in farm size, frequency of data sampling, and kind of visualisation results. The following categories of estimated data help quantify potential ranges of data volumes to be managed by a virtual farm:

Data Sources - Satellite, airborne, in situ:

Satellite Remote Sensing (Sample Sources to Estimate Data Volumes):

Whole data:

SPOT 60 × 60 km scene = 3 – 4 GB @ 1m multi-spectral 16 days (use as model for farm-specific application)

Subset data: 1000 acre farm from a 60x60km SPOT scene = .001 × 4 GB = .004 GB SPOT4 60 × 60 km scene = 36MB @ 10m panchromatic 16 days (use for farm and locality applications)

Subset data: .1% x 36 MB = .0036 MB

ASTER 60 × 60 km scene = 118 MB @ 15m 16 days (use for large farm and local/regional information input)

MODIS 10 × 10 degree scene = 139-680 MB @ 250m daily (use for regional context perspective)

Airborne Remote Sensing:

Example taken from AISA (Airborne Imaging Spectrometer for Applications)

1000 ft altitude = 1m pixel, 28 channels, swath 286 pixels wide. Variable-size data products

owing to flight path and duration

~2.8 GB coverage/scan

4000 ft altitude = 4m pixel, 28 channels, variable size data products

~.8 GB for coverage/scan

In Situ Sources:

Mobile mounted—planters, harvesters, applicators, etc. = ~1 MB/hour operation

Static / configurable (densities variable)—data loggers = 4 B/hour × (50/acre * 1000) = 200 KB/hour

Micro-electro-mechanical systems (MEMS) = 10 KB/reading × (50000 – 1000000) = 500 KB –10 MB/aggregate reading (MEMS could also be deployable as microclimate sensors)

Modeling Results, Visualisation, and Simulation Sources:

Weather Information Sources:

Microclimate zones (farm-situated)—monitored with the *in situ* sensors described above. These would provide the highest resolution weather data.

Local weather information—extracted from weather/climate reporting services (i.e., local news services). These sources would involve lower resolutions appropriate for monitoring changing conditions affecting the locale for the farm (e.g., hourly temperature, precipitation, cloudiness, forecasts)

Regional weather and climate histories and forecasts—extracted from state, federal and institutional modeling services. These sources provide the big picture at lower resolutions but yield longer-range forecasts and also histories of trends and patterns affecting the farm's region.

Forecast data feeds—from weather service providers. These could be a subset or could be scaled for farm concerns from master output models and visualisations. Data files for these might be 100MB to 1GB for high-resolution 3-D interactive visualisations. Visualisation files that are static viewable animations would be smaller (5 –30 MB MPEG for example).

An active zoom-able visualisation "window" on weather would involve a fusion of *in situ* sensor data and local/regional weather models/visualisations. Downloaded modeling data or preprocessed visualisation files would be transformed and scaled within the parameters of the virtual farm. These data would have value as temporary and persistent stored resources. For example, a three day moving "window" on weather updated hourly from external sources can be maintained while selected aspects of this view can be stored into a history file. Near-term storage of weather visualisation files for a virtual farm could involve 100-500 MB for a three-day window (past, present, future forecast), depending on whether they are 2-D or 3-D. Compression techniques could reduce the daily files for history files but accumulated visualisation data require large storage capacities.

Other:

In addition to weather data, the virtual farm will need to manage the data files produced as a result of selected modeling, visualisation, and simulations. These secondary data products (based on image data, metadata, or statistics) can be output from value-added or scientific modeling algorithms designed to support precision farming scaled to the dimensions of the farm. Data files like these constitute externally preprocessed data or inputs that are used in conjunction with virtual farm software functions. Examples include custom models of local drainage patterns, ecosystems and habitats, soils, ground water and aquifer, seed performance histories, agriculture techniques, and economic models. Speculative estimates for these kinds of data management requirements are made with regard to interactive versus static categorisation (see Table 1). The largest files will be for interactive preprocessed visualisation data representing multi-dimensional temporal, spatial, and parametric information. Persistence and high-resolution requirements will swell the data volume over time starting in the gigabyte range. Stored preprocessed data files such as these support powerful interactive information visualisations, simulations for "what-if" investigations, and modeling/analysis programmes to study histories and make projections regarding farm operations. Integrated external and internal preprocessed data assets represent extensions of farm-centric information to support specific operations (e.g., land management, profit maps, nutrient profiles, soil maps, chemical application maps, plantings, crop growth health, performance, crop yields, disease profiles, pest damage, harvest profiles, regulation compliance). Preprocessed data generated externally and internally (within the virtual farm) represents a significant data management requirement, one that an IA can address. For instance, cumulative, persistent storage of these data could represent hundreds of terabytes that a virtual farm could manage with virtual storage services offered by an IA.

Table 1: Estimated Data Consumption for 1000 Acre Farm

Category	Daily	Monthly	Annual	Subset
Satellite				
High 1m	16 day pass	8GB	96GB	
Med. 10m	""	72MB	.864GB	344.496GB x .1
Med. 15m	""	236MB	2.832GB	reduction = 34.45GB
Low 250m	680MB	20.4GB	244.8GB	
Airborne				
High 1m	2.8GB	*Planting:*		
Med 4m	.8GB	2 x 2/mo. = 14.4GB		
		Cultivating;		
		4 x 4/mo. = 57.6GB	*Total* 86.4GB	
		Harvesting:		
		2 x 2/mo. = 14.4GB		
		Winter 4mo: 0		
In Situ				
Mobile	6MB	*Planting:*		
Static	4.8MB	84, 48, MB	Mobile &	
		Cultivating:	static Total	
		400, 48 MB	684MB	
		Harvesting:		
		200, 48MB		
Configurable MEMS	240MB	7.2GB	86.4GB	
Visualisations	3 day window	30 day history	cumulative	
*Weather***	100-500MB	5GB	60GB	3-day window +
Interactive	5-30MB	900MB	10.8GB	compressed annual
Static			60GB	= 67GB
Other ***				
Totals			648.8GB	274.9GB

Extrapolating estimated annual data consumption for single 1000 acre farms to the 600,000 acres of the Central Valley agriculture zone in California yields 275 GB x 600 = 165,000 GB. This represents a potential distribution of 165 TB managed data for just one agricultural region. Persistence requirements may decrease at the virtual farm provided the intelligent archive provides faithful historical data that can be requested on-demand. Hence requirements for data persistence will be high for archiving services and rolling deletions strategies may be employable at the consumer end (e.g. virtual farm) though the option to retain history data is always available to it. Further extrapolation of this model to the entire United States (and to international applications) suggests intelligent data management, data understanding, distribution strategies, and information transformation services for agriculture alone will be in high demand.

REFERENCES

Brozovic, Nicholas, J. Carey, D Sunding, "How Big are Farms in the Westlands?", July 2001, http://are.berkeley.edu/~brozovic/econ/westland/westsize.pdf

Harberts, R., et al, "Intelligent Archive visionary Use Case: Advance Weather Prediction", September, 2003, http://daac.gsfc.nasa.gov/IDA/

Lockheed Martin Advanced Concepts Center and Rational Software Corporation, Succeeding with the Booch and OMT Methods: A Practical Approach, Addison-Wesley, 1996

Quatrani, Terry, Grady Booch, "Visual Modeling with Rational Rose and UML" 21-39, Addison-Wesley, 1998

Ramapriyan, H.K., G. McConaughy, C. Lynnes, S. Kempler, K. McDonald, R. Harberts, L. Roelofs, P. Baker, 2002. "Conceptual Study of Intelligent Archives of the Future", Report prepared for the Intelligent Data Understanding programme, http://daac.gsfc.nasa.govIDA/IA_report_8-27-02_baseline.pdf.

16

Use of GIS in Agro-ecological Zoning

Agro-ecological zoning (AEZ) is one of the most important bases for agricultural developmental planning because survival and failure of particular land use or farming system in a given region heavily relies on careful assessment of agro-climatic resources. A practical zoning approach thus arises because climate represented by thermal and moisture regimes forms small geographic areas, resulting in a variable mosaic of specialised areas, capable of supporting varied land use systems. The approach is used to categorise agroclimatically uniform geographical areas for agricultural developmental planning and other interventions. A framework of agro-ecological zoning describing concepts, methods and procedures was conceptualised for the first time by FAO. Agro-ecological zoning refers to the division of an area of land into land resource mapping units, having unique combination of landform, soil and climatic characteristics and or land cover having a specific range of potentials and constraints for land use.

The particular parameters used in the definition focus attention on the climatic and edaphic requirements of crop and on the management systems under which the crops are grown. Each zone has a similar combination of constraints and potentials for land use and serves as a focus for the targeting of recommendations designed to improve the existing land use situation, either through increasing production or by limiting land degradation. The addition of further layers of information on such factors as land tenure, land availability, nutritional requirement of human and livestock populations, infrastructure and costs and prices, has enabled the development of more advanced applications in natural resource analysis and land use planning. AEZ can be regarded as a set of core applications, leading to an assessment of land suitability and potential productivity. An output of AEZ studies includes maps showing agro-ecological zones and land suitability, quantitative estimates on potential crop yields and production. Such information provides the basis for advanced applications such as land degradation assessment, livestock productivity modeling, population support capacity assessment and land use optimisation modeling.

IMPORTANCE OF AEZ IN SUSTAINABLE AGRICULTURAL DEVELOPMENT PLANNING

The ability of the world's natural resources to provide for the needs of its growing population is a fundamental issue for the international community. World's population is increasing rapidly and at the same time, essential natural resources, such as land and water, are declining both in quantity and quality due to such factors as competition with industrial and urban demands. The basic problem is that limits to the productive capacity of land resources are set by climate, soil and land forms condition. In this context AEZ can be regarded as a set of applications, leading to an assessment of land suitability and potential productivity in terms of climate, soil and land forms condition.

Agro-Ecological Zoning (AEZ) is one of the most important bases of sustainable agricultural development planning of a region. It is applicable in micro or local level planning mainly for rainfed agriculture. It assesses basically the yield potentialities of various crop conditions; evolves future plan of action involving crop diversification; determines suitability of different crops for optimising land use, disseminates research results and agro-technology. As a result sustainable agricultural development planning is increasingly being based on agro-ecological zones. In this process agro climate zoning has become very popular. The initial focus of the FAO agro-ecological zoning system was to assess the suitability of different types of land use for selected land uses. It is an important starting point for selected land use planning with an overview of the whole region. It diagnoses the present situation with regard to farming and land use by categorising, describing and analyzing, farming systems components.

DEFINITIONS

1. *Agro-Ecological Zoning (AEZ)* refers to the division of an area of land into smaller units, which have similar characteristics related to land suitability potential production and environmental impact.
2. *Agro-Ecological Zone* is a land resource mapping unit, having a unique combination of land form, soil and climatic characteristics and/or land cover having a specific range of potentials and constraints for land use.
3. *Agro-Ecological Cell (AEC)* is defined by a unique combination of land form, soil and climatic characteristics

TRADITIONAL APPROACH

Several attempts have been made to classify the land area into climatic regions. Many of the earlier efforts to delineate agro-climates used manual overlay of isolines representing either potential evapotranspiration or temperature or their combinations and superimposed on soil resource maps. Carter divided India into six climatic regions, ranging from arid to per humid, based on the criteria of Thornthwaite system of climate classification. Sehgal *et al.* prepared a computerised bio-climatic map of NW India, based on the criteria of dry month. Krishnan delineated 40 soil climatic zones based on major soil types and moisture index. Murthy and Pandey brought out a 8 agro-ecological region map of India on the basis of physiography, climate, soils and agricultural regions. The approach depicts a good beginning of agro-ecological zoning in the country, but it suffers from several limitations due to over generalisations such as grouping together the areas having different physiography, temperature and soil in zone. Subramanian based on the data of 160 meteorological stations in the country and using the concept of moisture adequacy index, delineated 29 agro-ecological zones with the possible 36 combinations of IMA and dominated soil groups as per FAO/UNESCO Soil Map.The planning commission, as a result of mid-term appraisal of the planning targets of VII Plan, divided the country into 15 broad agro-climatic zones based on physiography and climate.

New Tools For AEZ

Modern tools such as satellite remote sensing and GIS have been providing newer dimensions to effectively monitor and manage natural resources. It has been well conceived that remote sensing and GIS have great role to play in agro-ecological zoning for sustainable development due to multi-stage character of the comprehensive approach to agro-ecological zoning. Several approaches of AEZ in past involved manual integration of agro-climatic and other natural resource data. As a result, large amount of agro-ecological data could not be handled easily and aggregation was required at an early stage in the analysis. This led to loss of information on spatial variability. On the other hand, GIS technology is very useful for automated logical integration of bio-climate, terrain and soil resource inventory information. The system is capable of containing all data required to solve resource management problems. Topographic maps, land resource map and contour map having physiographic, geographic and bio-climatic information forms primary input for GIS for agroecological zoning activities. The system also facilitates the enlargement of a particular geographic pocket to render more details on retrieval. After collecting the basic data on zonal resource information, the data can be manipulated to create relevant profiles of applied use that can be retrieved on demand. A zonal database can also be integrated with non-geographic information such as socioeconomic data, which is relevant for making decision on development priority interventions about the sustainable management of zonal resources. Remote sensing provides digital or hard copy data base information on natural resources. This information can be stored and retrieved as and when required and also data can be classified and aggregated for any number of planning exercises. This AEZ concept involves the representation of land in layers of spatial information and combination of layers of spatial information using geographic information system (GIS).

STRUCTURE OF AEZ

The structure of AEZ includes the comprehensive framework for the appraisal and planning of land resources. The nature of analysis, which involves the combination of layers of spatial information to define zones, lends itself to application of a GIS. The major requirements of computerised GIS for an activity like agro-ecological zoning and zonal resource information of mountain areas are topographic maps, land resource maps and contour maps. These maps, containing physiographic, geographic, and bio-climatic information form primary inputs. Various outputs are generated in both tabular and map forms. Till date, good progress has been made in developing GIS based tools for land resources planning, management and monitoring at different scales.

ELEMENTS OF AEZ

The essential elements of the core applications of AEZ include :

1. Land resource inventory, comprising
   (a) Analysis of length of growing period or moisture availability index
   (b) Defining thermal zones
   (c) Compilation of climatic resource inventory
   (d) Compilation of soil and landform resource inventory
   (e) Compilation of present land use inventory
   (f) Combination of above to make land resource inventory based on agroecological zones or agro-ecological cells.
2. Inventory of land utilisation types and crop requirements
3. Land suitability evaluation, including
   (a) Potential productivity computations
   (b) Matching of constraints and requirements

A CASE STUDY ON AGRO-ECOLOGICAL ZONING IN MOUNTAIN ECOSYSTEM

Sustainable development of mountain regions is a challenging task because the areas have highly diverse and fragile ecosystems. One of the most striking characteristics of mountains is their spatial variability. This makes the planning of the use of natural resources in the mountains more complex than any other area. In view of this, the present study was conducted in part of Kumaon Himalayas (latitudes 28°45′ to 30°00′N, longitude 78°45′ to 80°15′) to demonstrate the use of remote sensing and GIS as a tools for agro-ecological zoning with mountain perspective. The methodology used in this study is outlined in the flow diagram in Figure 1.

Climate

Long term (approximately 10 years) monthly maximum and minimum temperatures were collected from six meteorological stations falling in the western Himalayas.

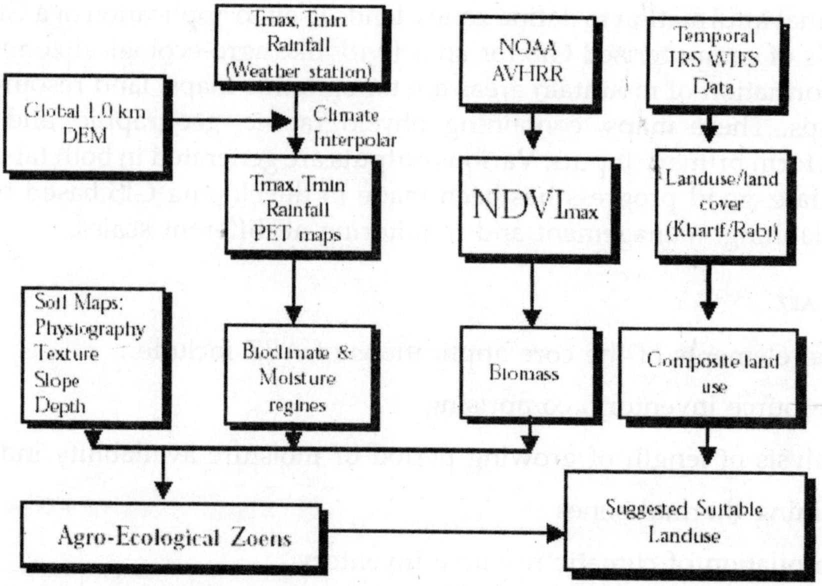

Figure 1 : Flow diagram of the method for agro-ecological zoning Soil Maps:

However, monthly rainfall data for eight years were collected from eleven rain gauge stations within the study area. Digital elevation data with one-kilometer grid size was taken from Global Digital Elevation Model (USGS GDEM) and geometrically registered in UTM projection. Validation and accuracy check for digital elevation data derived from Global DEM was done with spot height observation from Survey of India toposheets on 1:250000 scale. A close agreement was observed between spot height observations and Global DEM based elevation data (R^2 = 0.98). Long term monthly and annual averages of mean temperatures of six meteorological stations were regressed against corresponding elevation using MICROSTA statistical software (Table 1).

Table 1. Mean Temperature v/s Elevation Relationship

Month	Regression Equation	R^2
January	y = - 0.0031 x + 14.106	0.89
February	y = - 0.004 x + 17.012	0.95
March	y = - 0.0047 x + 22.122	0.97
April	y = - 0.0051 x + 27.95	0.96
May	y = - 0.0057 x + 31.485	0.98
June	y = - 0.0059 x + 31.925	0.96
July	y = - 0.0058 x + 30.642	0.99
August	y = - 0.0054 x + 29.755	0.98
September	y = - 0.0053 x+ 28.855	0.99
October	y = - 0.0044 x + 25.133	0.95
November	y = - 0.003 x + 19.284	0.89
December	y = - 0.0018 x + 15.045	079

A good agreement was also observed between annual mean temperature and elevation (annual mean temperature = 24.443 – 0045 × elevation, R^2 = 0.97). Similarly long term average annual rainfall recorded at different rain gauge stations were regressed against elevation for developing rainfall—elevation relationship (annual rainfall = 515.1 + 0.3843 × elevation, R^2 = 0.75) representing the region under study. The empirical relations thus developed were used to utilise inherent spatial quality of digital elevation model in GIS environment for depicting spatial variation in normal monthly and annual mean temperatures as well as annual rainfall condition over Kumaon region. These monthly spatial distribution of mean temperatures were used for computation of spatial potential evapotranspiration (PET) based on Thornthwaite (1948) as:

$$\text{PET} = 1.6\ [(10\ T_{ijk}/I_{jk})^{ajk}]$$

where,

$$I_{jk} = \sum_{t=\text{Jan}}^{\text{Dec}} (T_{ijk}/5)^{1.514}$$

T = Mean air temperature (°C)

i = Month of a year (i = Jan, Feb, ...Dec)

j = Pixel value of ith row

k = Pixel value of jth column

a_{jk} =67.5*10^{-8} (I_{jk}) 3 – 7.71*10 – 5*(I_{jk}) 2 + 0.01792 (I_{jk}) + 0.4923

These monthly potential evapotranspiration were summed over twelve months in a year to obtain spatial distribution of annual PET for use in computation of moisture index (MI). The moisture index with positive and negative values indicates moist or dry climate and seasonal variation of effective moisture. The revised moisture index of Thornwaite and Mather method based on annual rainfall and potential evapotranspiration was calculated as,

$$\text{MI} = [(P - \text{PET})/\text{PET}] * 100$$

where,

MI = Moisture Index

P = Rainfall (mm)

PET = Potential Evapotranspiration (mm)

This information on moisture index is vital to congenial biotic environment and has been used to classify the climatic types. To arrive at homogeneous zones, the limit of moisture index of the various climatic types was scaled as shown in Table 2.

Table 2. Scaling of moisture index into different agro-climate types

Climatic type	Moisture Index	Symbol
Semi arid	–66.7 to –33.3	D
Dry sub-humid	–33.3 to 0	C1
Moist sub-humid	0 to 20	C2
Humid (4 classes)	20 to 100	B1, B2, B3 and B4
Per-humid	> 100	A

Vegetation Biomass variability

Using NDVI to estimate standing green biomass proved to be a reliable source of biomass data. The existing functional relationship between monthly cutting of green dry matter (DM) and maximum normalised difference vegetation index (NDVI) derived from NOAA AVHRR (Advanced Very High Resolution Radiometer) was used. Monthly NDVI images of NOAA AVHRR during year 1999 were used to derive the maximum NDVI in a year for estimating biomass or dry matter (DM) as per equation.

$$DM = (1.615 * NDVI_{max})^{1.318} \; ; \; R^2 = 0.90$$

$$DM = \text{Dry matter}$$

$$NDVI_{max} = \text{Maximum NDVI in a year.}$$

Land Use/land Cover

Land use/land cover was derived from coarse resolution IRS WiFS satellite data. Ground truth was collected through integrated use of previous year IRS LISS III hard copy (1:250,000 scale) satellite data, Survey of India (SOI) toposheets and handheld Global Positioning System (GPS). Combination of satellite data acquired during *kharif* and *rabi* easons were digitally classified to land use/land cover information classes for *kharif* and *rabi*, respectively using MXL classifier. Agricultural land use classes in *rabi* season were refined with respect to land use information in *kharif* season and crop calendar of major crops cultivated in the region. Finally, classified land use /land cover information of *kharif* and *rabi* seasons were logically integrated on a pixel by pixel basis in raster GIS for deriving cropping system or composite land use/land cover.

Soils

Forty-six soil mapping unit along with their characteristics such as depth, texture, slope, erosion status, and drainage condition for the study area were obtained from soil map (1:250000 scale) prepared by National Bureau of Soil Survey and Land Use Planning, India. These spatial and non spatial data on soils were converted into digital soil resource databases for AEZ. In general terms, coarse loamy and skeletal soils are in side slope and top of Lesser Himalayas and piedmont plain while fine textured soils mainly found in alluvial plain and fluvial valleys. The soils are mostly moderately deep to deep, however, approximately 25 % of the area having shallow soils were mainly found in Lesser Himalayas and piedmont plain.

Delineation of agro-climatic zones

Agro-climatic zones are of paramount importance for defining or delineation of agro-ecological zoning for sustainable use of land resources. The essential elements in demarcating or defining of an agro-climatic zone are bioclimates based on thermal regimes, moisture regime and biomass variability 226 Remote Sensing and GIS Application in Agro-Ecological Zoning. Thermal regimes indicate to the amount of heat available for plant growth and development during the growing period. Moisture regimes represent water availability for crop production and hence they are the vital to classify the agro-climatic zones. Nine different agro-climatic zones were delineated by GIS aided integration of thermal regimes, moisture regimes and biomass map layers.

Delineation of Agro-ecological Zones and Sub-zones

Thirty agro-ecological sub zones (AEZIa to AEZXIIc) were delineated in the study area by GIS aided integration of nine agro-climatic and eight agro-edaphic zones. Results revealed that the majority of area falls in XIa whereas minimum area is occupied by Via. The other larger sub-zones are Ib and VIIIa, respectively. AEZIc and AEZIIc are the most potential zones for intensive cropping with possibility of two and three crops such as paddy—wheat or sugarcane—wheat. Further sub-ecological zones are reclassified by GIS aided integration in which subzones are clubbed into main agro-ecological zones. Therefore, twelve agroecological zones (AEZI to AEZXII) are identified in the study area.

REFERENCES

Carter, D.B. 1954. "Climates of Africa and according to Thornthwaite (1948) Classification". John Hopkins University Publication (in) Climatology, 7(14).

FAO. 1976 "A framework for land evaluation". Food and Agricultural Organisation, *Soils Bulletin* 32, Rome, Italy.

Krishnan, A. 1988. "Delineation of soil climatic zones of India and its practical applications in Agriculture". Fertiliser News, 33 (4).

Mavi, H.S. 1984. "Introduction to Agrometeorology", IInd edition. Raju Primlani for Oxford & IBH Publishers Co. Pvt. Ltd.

Sehgal, J.L., Vernemmen, C. and Tavernier, R. 1987. "Agro-climatic Environments and Moisture Regimes in NW India—Their Applications in Soils and Crop Growth". *Research Bull.* NBSS Publ. 17, Nagpur, India, NBSS & LUP.

Sehgal, J.L., Mandal, D.K., Mandal, C. and Vadivelu, S. 1992. "Agro-ecological Region of India". NBSS & LUP (ICAR) Publication 24, Nagpur.

Subramanian, A.R. 1983. "Agro-ecological Zones of India". *Arch. Met. Geoph. Biocl. Ser.* B32; 329-333.

17

Modelling of Crop Growth

Crop is defined as an "Aggregation of individual plant species grown in a unit area for economic purpose". Growth is defined as an "Irreversible increase in size and volume and is the consequence of differentiation and distribution occurring in the plant". Simulation is defined as "Reproducing the essence of a system without reproducing the system itself ". In simulation the essential characteristics of the system are reproduced in a model, which is then studied in an abbreviated time scale. A model is a schematic representation of the conception of a system or an act of mimicry or a set of equations, which represents the behaviour of a system. Also, a model is "A representation of an object, system or idea in some form other than that of the entity itself ". Its purpose is usually to aid in explaining, understanding or improving performance of a system. A model is, by definition "A simplified version of a part of reality, not a one to one copy". This simplification makes models useful because it offers a comprehensive description of a problem situation. However, the simplification is, at the same time, the greatest drawback of the process. It is a difficult task to produce a comprehensible, operational representation of a part of reality, which grasps the essential elements and mechanisms of that real world system and even more demanding, when the complex systems encountered in environmental management.

The Earth's land resources are finite, whereas the number of people that the land must support continues to grow rapidly. This creates a major problem for agriculture. The production (productivity) must be increased to meet rapidly growing demands while natural resources must be protected. New agricultural research is needed to supply information to farmers, policy makers and other decision makers on how to accomplish sustainable agriculture over the wide variations in climate around the world. In this direction explanation and prediction of growth of managed and natural ecosystems in response to climate and soil-related factors are increasingly important as objectives of science. Quantitative prediction of complex systems, however, depends on integrating information through levels of organisation, and the principal approach for that is through the construction of statistical and simulation models. Simulation of system's use and balance of carbon, beginning with the input of carbon from canopy assimilation forms the essential core of most simulations that deal with the growth of vegetation.

Systems are webs or cycles of interacting components. Change in one component of a system produces changes in other components because of the interactions. For example, a change in weather to warm and humid may lead to the more rapid development of a plant disease, a loss in yield of a crop, and consequent financial adversity for individual farmers and so for the people of a region. Most natural systems are complex. Many do not have boundaries. The bio-system is comprised of a complex interaction among the soil, the atmosphere, and the plants that live in it. A chance alteration of one element may yield both desirable and undesirable consequences. Minimising the undesirable, while reaching the desired end result is the principle aim of the agrometerologist. In any engineering work related to agricultural meteorology the use of mathematical modeling is essential. Of the different modeling techniques, mathematical modeling enables one to predict the behaviour of design while keeping the expense at a minimum. Agricultural systems are basically modified ecosystems. Managing these systems is very difficult. These systems are influenced by the weather both in length and breadth. So, these have to be managed through systems models which are possible only through classical engineering expertise.

TYPES OF MODELS

Depending upon the purpose for which it is designed the models are classified into different groups or types. Of them a few are :

1. *Statistical models:* These models express the relationship between yield or yield components and weather parameters. In these models relationships are measured in a system using statistical techniques. Example: Step down regressions, correlation, etc.
2. *Mechanistic models:* These models explain not only the relationship between weather parameters and yield, but also the mechanism of these models (explains the relationship of influencing dependent variables). These models are based on physical selection.
3. *Deterministic models:* These models estimate the exact value of the yield or dependent variable. These models also have defined coefficients.
4. *Stochastic models:* A probability element is attached to each output. For each set of inputs different outputs are given alongwith probabilities. These models define yield or state of dependent variable at a given rate.
5. *Dynamic models:* Time is included as a variable. Both dependent and independent variables are having values which remain constant over a given period of time.

6. *Static:* Time is not included as a variables. Dependent and independent variables having values remain constant over a given period of time.

7. *Simulation models:* Computer models, in general, are a mathematical representation of a real world system. One of the main goals of crop simulation models is to estimate agricultural production as a function of weather and soil conditions as well as crop management. These models use one or more sets of differential equations, and calculate both rate and state variables over time, normally from planting until harvest maturity or final harvest.

8. *Descriptive model:* A descriptive model defines the behaviour of a system in a simple manner. The model reflects little or none of the mechanisms that are the causes of phenomena. But, consists of one or more mathematical equations. An example of such an equation is the one derived from successively measured weights of a crop. The equation is helpful to determine quickly the weight of the crop where no observation was made.

9. *Explanatory model:* This consists of quantitative description of the mechanisms and processes that cause the behaviour of the system. To create this model, a system is analyzed and its processes and mechanisms are quantified separately. The model is built by integrating these descriptions for the entire system. It contains descriptions of distinct processes such as leaf area expansion, tiller production, etc. Crop growth is a consequence of these processes.

WEATHER DATA FOR MODELING

The national meteorological organisations provide weather data for crop modeling purposes through observatories across the globe. In many European countries weather records are available for over 50 years. In crop modeling the use of meteorological data has assumed a paramount importance. There is a need for high precision and accuracy of the data. The data obtained from surface observatories has proved to be excellent. It gained the confidence of the people across the globe for decades. These data are being used daily by people from all walks of life. But, the automated stations are yet to gain popularity in the under developed and developing countries. There is a huge gap between the old time surface observatories and present generation of automated stations with reference to measurement of rainfall. The principles involved in the construction and working of different sensors for measuring rainfall are not commonly followed in automated stations across the globe. As of now, solar radiation, temperature and precipitation are used as inputs in DSSAT.

Weather as an Input in Models

In crop modeling weather is used as an input. The available data ranges from one second to one month at different sites where crop-modeling work in the world is going on. Different curve fitting techniques, interpolation, extrapolation functions etc., are being followed to use weather data in the model operation. Agrometeorological variables are especially subject to variations in space. It is reported that, as of now, anything beyond daily data proved unworthy as they are either over-estimating or under-estimating the yield in simulation. Stochastic weather models can be used as random number generators whose input resembles the weather data to which they have been fit. These models are convenient and computationally fast, and are useful in a number of applications where the observed climate record is inadequate with respect to length, completeness, or spatial coverage. These applications include simulation of crop growth, development and impacts of climate change.

In 1995 JW Jones and Thornton described a procedure to link a third-order Markov Rainfall model to interpolated monthly mean climate surfaces. The constructed surfaces were used to generate daily weather data (rainfall and solar radiation). These are being used for purposes of system characterisation and to drive a wide variety of crop and live stock production and ecosystem models. The present generation of crop simulation models particularly DSSAT suit of models have proved their superiority over analytical, statistical, empirical, combination of two or all etc., models so far available. In the earliest crop simulation models only photosynthesis and carbon balance were simulated. Other processes such as vegetative and reproductive development, plant water balance, micronutrients, pest and disease, etc., are not accounted for as the statistical models use correlative approach and make large area yield prediction and only final yield data are correlated with the regional mean weather variables. This approach has slowly been replaced by the present simulation models by these DSSAT models. When many inputs are added in future the models become more complex. The modelers who attempt to obtain input parameters required to add these inputs look at weather as their primary concern. They may have to adjust to the situation where they develop capsules with the scale level at which the input data on weather are available.

Role of Weather in Decision Making

Decisions based solely upon mean climatic data are likely to be of limited use for at least two reasons. The first is concerned with definition of success and the second with averaging and time scale. In planning and analyzing agricultural systems it is essential not only to consider variability, but also to think of it in terms directly relevant to components of the system. Such analyses may be relatively straightforward probabilistic analyses of particular events, such as the start of cropping seasons in West Africa and India. The principal effects of weather on crop growth and development are well understood and are predictable. Crop simulation models can predict responses to large variations in weather. At every point of application weather data are the most important input. The main goal of most applications of crop models is to predict commercial output (Grain yield, fruits, root, biomass for fodder etc.). In general the management applications of crop simulation models can be defined as: (1) strategic applications (crop models are run prior to planting), (2) practical applications (crop models are run prior to and during crop growth) and (3) forecasting applications (models are run to predict yield both prior to and during crop growth).

Crop simulation models are used in USA and in Europe by farmers, private agencies, and policy makers to a greater extent for decision making. Under Indian and African climatic conditions these applications have an excellent role to play. The reasons being the dependence on monsoon rains for all agricultural operations in India and the frequent dry spells and scanty rainfall in crop growing areas in Africa. Once the arrival of monsoon is delayed the policy makers and agricultural scientists in India are under tremendous pressure. They need to go for contingency plans. These models enable to evaluate alternative management strategies, quickly, effectively and at no/low cost. To account for the interaction of the management scenarios with weather conditions and the risk associated with unpredictable weather, the simulations are conducted for at least 20-30 different weather seasons or weather years. If available, the historical weather data, and if not weather generators are used presently. The assumption is that these historical data will represent the variability of the weather conditions in future. Weather also plays a key role as input for long-term crop rotation and crop sequencing simulations.

CLIMATE CHANGE AND CROP MODELING

Climate Change

Climate change is defined as "Any long term substantial deviation from present climate because of variations in weather and climatic elements".

The Causes of Climate Change

1. The natural causes like changes in earth revolution, changes in area of continents, variations in solar system, etc.
2. Due to human activities the concentrations of carbon dioxide and certain other harmful atmospheric gases have been increasing. The present level of carbon dioxide is 325 ppm and it is expected to reach 700 ppm by the end of this century, because of the present trend of burning forests, grasslands and fossil fuels. Few models predicted an increase in average temperature of 2.3 to 4.6°C and precipitation per day from 10 to 32 per cent in India.

Green House Effect

The effect because of which the earth is warmed more than expected due to the presence of atmospheric gases like carbon dioxide, methane and other tropospheric gases. The shortwave radiation can pass through the atmosphere easily, but, the resultant outgoing terrestrial radiation can not escape because atmosphere is opaque to this radiation and this acts to conserve heat which rises temperature.

Effects of Climate Change

1. The increased concentration of carbon dioxide and other green house gases are expected to increase the temperature of earth.
2. Crop production is highly dependent on variation in weather and therefore any change in global climate will have major effects on crop yields and productivity.
3. Elevated temperature and carbon dioxide affects the biological processes like respiration, photosynthesis, plant growth, reproduction, water use etc. In case of rice increased carbon dioxide levels results in larger number of tillers, greater biomass and grain yield. Similarly, in groundnut increased carbon dioxide levels results in greater biomass and pod yields.
4. However, in tropics and sub-tropics the possible increase in temperatures may offset the beneficial effects of carbon dioxide and results in significant yield losses and water requirements.

Proper understanding of the effects of climate change helps scientists to guide farmers to make crop management decisions such as selection of crops, cultivars, sowing dates and irrigation scheduling to minimise the risks.

Role of Climate Change in Crop Modeling

In recent years there has been a growing concern that changes in climate will lead to significant damage to both market and non-market sectors. The climate change will have a negative effect in many countries. But farmers adaptation to climate change-through changes in farming practices, cropping patterns, and use of new technologies will help to ease the impact. The variability of our climate and especially the associated weather extremes is currently one of the concerns of the scientific as well as general community. The application of crop models to study the potential impact of climate change and climate variability provides a direct link between models, agrometeorology and the concerns of the society. Cropgro (DSSAT) is one of the first packages that modified weather simulation generators/or introduced a package to evaluate the performance of models for climate change situations. Irrespective of the limitations of GCMs it would be in the larger interest of farming community of the world that these DSSAT modelers look at GCMs for more accurate and acceptable weather generators for use in models. This will help in finding solutions to crop production under climate changes conditions, especially in underdeveloped and developing countries.

FUTURE ISSUES RELATED TO WEATHER ON CROP MODELING

For any application of a crop model weather data is an essential input and it continues to play a key role. So:

1. There is an urgent need to develop standards for weather station equipment and sensors installation and maintenance.

2. It is also important that a uniform file format is defined for storage and distribution of weather data, so that they can easily be exchanged among agrometeorologists, crop modelers and others working in climate and weather aspects across the globe.
3. Easy access to weather data, preferably through the internet and the world wide web, will be critical for the application of crop models for yield forecasting and tactical decision making.
4. Previously one of the limitations of the current crop simulation models was that they can only simulate crop yield for a particular site. At this site weather (soil and management) data also must be available. It is a known fact that the weather data (and all these other details) are not available at all locations where crops are grown. To solve these problems the Geographical Information System (GIS) approach has opened up a whole field of crop modeling applications at spatial scale. From the field level for site-specific management to the regional level for productivity analysis and food security the role of GIS is going to be tremendous.

APPLICATIONS AND USES OF CROP GROWTH MODELS IN GRICULTURAL METEOROLOGY

The crop growth models are being developed to meet the demands under the following situations in agricultural meteorology.

1. When the farmers have the difficult task of managing their crops on poor soils in harsh and risky climates.
2. When scientists and research managers need tools that can assist them in taking an integrated approach to finding solutions in the complex problem of weather, soil and crop management.
3. When policy makers and administrators need simple tools that can assist them in policy management in agricultural meteorology.

On Farm Decision-making and Agronomic Management

The models allow evaluation of one or more options that are available with respect to one or more agronomic management decisions like:

1. Determine optimum planting date.
2. Determine best choice of cultivars.
3. Evaluate weather risk.
4. Investment decisions.

The crop growth models can be used to predict crop performance in regions where the crop has not been grown before or not grown under optimal conditions. Such applications are of value for regional development and agricultural planning in developing countries. A model can calculate probabilities of grain yield levels for a given soil type based on rainfall. Investment decisions like purchase of irrigation systems can be taken with an eye on long term usage of the equipment thus acquired. Kiniry *et al.* showed that for maize, both simulated and measured mean yields with weeds are 86% of the weed-free yields.

Understanding of Research

In agro-meteorological research the crop models basically helps in:

1. Testing scientific hypothesis.
2. Highlight where information is missing.
3. Organising data.
4. Integrating across disciplines.
5. Assist in genetic improvement;
   (a) Evaluate optimum genetic traits for specific environments.
   (b) Evaluate cultivar stability under long term weather.

Penning de Vries emphasised that simulation models contribute to our understanding of the real system which in-turn helps to bridge areas and levels of knowledge. It is believed that in conversion of conceptual models into mathematical simulation models the agrometeorologists can understand the gaps in their knowledge. So, the interdisciplinary nature of simulation modeling efforts leads to increased research efficacy and improved research direction through direct feedback. In this direction de Wit and Goudriaan developed BAsic CROp growth Simulator (BACROS) which was used as a reference model for developing other models and as a basis for developing summary models. Also O Toole and Stockle described the potential of simulation models in assessing trait benefits of winter cereals and their capacity to survive and reproduce in stress-prone environment. Crop growth models have been used in plant breeding to simulate the effects of changes in the morphological and physiological characteristics of crops which aid in identification of ideotypes for different environments.

Policy management

The policy management is one very useful application of crop simulation models. The issues range from global (impacts of climate change on crops) to field level (effect of crop rotation on soil quality) issues. Thornton *et al.* showed that in Burkina Faso, crop simulation modeling using satellite and ground-based data could be used to estimate millet production for famine early warning which can allow policy makers the time they need to take appropriate steps to ameliorate the effects of global food shortages on vulnerable urban and rural populations. In Australia Meinke and Hammer found that when November-December SOI (Southern Oscillation Index) phase is positive, there is an 80% chance of exceeding average district yields. Conversely, in years when the November-December SOI phase is either negative or rapidly falling, there is only a 5% chance of exceeding average district yields, but a 95% chance of below average yields. This information allows the industry to adjust strategically for the expected volume of production. Crop models can be used to understand the effects of climate change such as:

1. Consequences of elevated carbon-dioxide, and
2. Changes in temperature and rainfall on crop development, growth and yield. Ultimately, the breeders can anticipate future requirements based on the climate change.

A FEW SUCCESSFULLY USED MODELS IN AGROMETEOROLOGY

Large scale evolution of computers since 1960 allowed to synthesise detailed knowledge on plant physiological processes in order to explain the functioning of crops as a whole. Insights into various processes were expressed using mathematical equations and integrated in simulation models. In the beginning, models were meant to increase the understanding of crop behaviour by explaining crop growth and development in terms of the understanding physiological mechanisms. Over the years new insights and different research questions motivated the further development of simulation models. In addition to their explanatory function, the applicability of well-tested models for extrapolation and prediction was quickly recognised and more applicationoriented models were developed. For instance demands for advisory systems for farmers and scenario studies for policy makers resulted in the evolution of models, geared towards tactical and strategic decision support respectively. Now, crop growth modeling and simulation have become accepted tools for agricultural research.

The de Wit school of models

In the sixties, the first attempt to model photosynthetic rates of crop canopies was made. The results obtained from this model were used among others, to estimate potential food production for some areas of the world and to provide indications for crop management and breeding. This was followed by the construction of an Elementary CROp growth Simulator (ELCROS) by de Wit *et al.*. This model included the static photosynthesis model and crop respiration was taken as a fixed fraction per day of the biomass, plus an amount proportional to the growth rate. In addition, a functional equilibrium between root and shoot growth was added. The introduction of micrometeorology in the models and quantification of canopy resistance to gas exchanges allowed the models to improve the simulation of transpiration and evolve into the BAsic CROp growth Simulator (BACROS).

IBSNAT and DSSAT Models

In many countries of the world, agriculture is the primary economic activity. Great numbers of the people depend on agriculture for their livelihood or to meet their daily needs, such as food. There is a continuous pressure to improve agricultural production due to staggering increase in human population. Agriculture is very much influenced by the prevailing weather and climate. The population increase is 2.1 per cent in India. This demands a systematic appraisal of climatic and soil resources to recast an effective land use plan. More than ever farmers across the globe want access to options such as the management options or new commercial crops. Often, the goal is to obtain higher yields from the crops that they have been growing for a long time. Also, while sustaining the yield levels they want to :

1. Substantially improve the income.
2. Reduce soil degradation.
3. Reduce dependence on off-farm inputs.
4. Exploit local market opportunities.
5. Farmers also need a facilitating environment in which;
   (a) Affordable credit is available.
   (b) Policies are conducive to judicious management of natural resources.
   (c) Costs and prices of production are stable.
6. Another key ingredient of a facilitating environment is information, such as :
   (a) An understanding of which options are available.
   (b) How these operate at farm level.
   (c) The impact on issues of their priority.

To meet the above requirements of resource poor farmers in the tropics and sub tropics IBSNAT (International Benchmark Sites Network for Agrotechnology Transfer) began in 1982. This was under a contract from the U.S. Agency for International Development to the University of Hawaii at Manoa, USA. IBSNAT was an attempt to demonstrate the effectiveness of understanding options through systems analysis and simulation for ultimate benefit of farm households across the globe. The purposes defined for the IBSNAT project by its technical advisory committee were to :

1. Understand ecosystem processes and mechanisms.
2. Synthesise from an understanding of processes and mechanisms, a capacity to predict outcomes.
3. Enable IBSNAT clientele to apply the predictive capability to control outcomes.

The models developed by IBSNAT were simply the means by which the knowledge scientists have and could be placed in the hands of users. In this regard, IBSNAT was a project on systems analysis and simulation as a way to provide users with options for change. In this project many research institutions, universities, and researchers across the globe spent enormous amount of time and resources and focused on:

1. Production of a "decision support system" capable of simulating the risks and consequences of alternative choices, through multi-institute and multidisciplinary approaches.
2. Definition of minimum amount of data required for running simulations and assessing outcomes.
3. Testing and application of the product on global agricultural problems requiring site-specific yield simulations.

The major product of IBSNAT was a Decision Support System for Agro-Technology Transfer (DSSAT). The network members lead by J.W. Jones, Gainesville, USA developed this. The DSSAT is being used as a research and teaching tool. As a research tool its role to derive recommendations concerning crop management and to investigate environmental and sustainability issues is unparalleled. The DSSAT products enable users to match the biological requirements of crops to the physical characteristics of land to provide them with management options for improved land use planning. The DSSAT is being used as a business tool to enhance profitability and to improve input marketing.

The traditional experimentation is time consuming and costly. So, systems analysis and simulation have an important role to play in fostering this understanding of options. The information science is rapidly changing. The computer technology is blossoming. So, DSSAT has the potential to reduce substantially the time and cost of field experimentation necessary for adequate evaluation of new cultivars and new management systems. Several crop growth and yield models built on a framework similar in structure were developed as part of DSSAT package. The package consists of: (1) data base management system for soil, weather, genetic coefficients, and management inputs, (2) Cropsimulation models, (3) series of utility programmes, (4) series of weather generation programmes, (5) strategy evaluation programme to evaluate options including choice of variety, planting date, plant population density, row spacing, soil type, irrigation, fertiliser application, initial conditions on yields, water stress in the vegetative or reproductive stages of development, and net returns.

Other Important Models

Crop models can be developed at various levels of complexity. The level of complexity required depends on the objective of the modeling exercise. The top-down approach to model design is appropriate for models aimed at yield prediction. In this approach, complexity is kept to a minimum by commencing with a simple framework and only incorporating additional phenomena or processes if they improve the predictive ability of the model. Sinclair, Muchow *et al.* and Hammer and Muchow have adopted this method in developing models of soybean, maize and sorghum respectively.

The EPIC, ALAMANC, CROPSYST, WOFOST, ADEL models are being successfully used to simulate maize crop growth and yield. The SORKAM, SorModel, SORGF as also ALMANAC models are being used to address specific tasks of sorghum crop management. CERES—pearl millet model, CROPSYST, PmModels are being used to study the suitability and yield simulation of pearl millet genotypes across the globe. Similarly, the two most common growth models used in application for cotton are the GOSSYM and COTONS models. On the same analogy the PNUTGRO for groundnut, CHIKPGRO for chick pea, WTGROWS for wheat, SOYGRO for soybean, QSUN for sunflower are in use to meet the requirements of farmers, scientists, decision makers, etc., at present. The APSIM, GROWIT added with several modules are being used in crop rotation, crop sequence and simulation studies involving perennial crops.

Under Indian sub-continent conditions, crop yield forecasting based on meteorological data is very important from several points of view. Using crop yield as dependent variable and weather factors as independent variables, empirical statistical models for predicting crop yield have been reported by Mall *et al.*. Also, Mall and Gupta reported a successful empirical statistical – yield weather model for wheat in the Varanasi district of Uttar Pradesh, India. Rajesh Kumar *et al.* developed a stepwise regression model which states that the pigeonpea yield variation by weather variables is upto 94 per cent in Varanasi District, U.P., India. Using a thermal time and phasic development model. Patel *et al.* found that pigeonpea phenophases depended upon the available heat units and accounted for 98 per cent of total variation in Anand, Gujarat State, India.

Elsewhere in the world there were studies, which predicted yields based on climatic factors using mathematical modeling. Factors such as water deficit, solar radiation, maximum and minimum temperatures which play a vital role at floral initiation were taken into consideration to construct these models. In Southern Malaysia Ong found a high correlation between yield of oil palm and rainfall and dry spells as also temperature and sunshine using a step-wise regression approach. Chow constructed a statistical model for predicting crude palm oil production with trend, season, rainfall, etc. Amissah-Arthur and Jagtap successfully assessed nitrogen requirements by maize across agro ecological zones in Nigeria using CERES-maize model. Hammer *et al.* using local weather and soil information correlated peanut yields with estimates from PEANUTGRO, a model in the CERES family and gave a regression with high coefficient ($r^2 = 0.93$) of variation. The construction of contemporary crop models entails the combination of many algorithms for physiological processes and impact of environmental factors on process rates. This clearly indicates that in the development of models and their application for solving problems at field level on agrometeorological aspects are given due weightage.

APPLICATION OF MODELS IN AGRICULTURAL METEOROLOGY PRECAUTIONS

Hunger demands food. Food grains come from agriculture, but it is riskprone. So research efforts are necessary. Unfortunately, funding for research worldwide is declining at an alarming rate. The agricultural scientists and planners are facing formidable challenges to ensure continued increases in agricultural productivity to meet the food grain requirements of burgeoning population across the globe. Nowadays, the traditional field experimentation is becoming very expensive. So, the works on development and use of crop growth models to answer strategic and tactical questions concerning agricultural planning as well as on-farm soil and crop management are essential. Although these models are useful in more than one way as detailed earlier in this lecture, much of the modeling uses and applications to date have been mainly in the area of research. The products of research need to be disseminated to the farmers. Otherwise the present works on crop modeling cannot sustain in the long run. So far, much of crop modeling applications in the literature are based on sole cropping and only in the recent past this work has been extended to intercrops, crop rotations, crop sequences, etc. The works on crop rotations shall be taken up on priority. If one considers the attention given to individual crops so far maize and cotton rank more than 50% of the work but there are many other crops that meet food requirements of the population in tropical and sub-tropical areas of the world. So, it is essential to understand fully the physical and physiological processes of these crops that govern their growth and yield under valuable soil and climatic conditions through crop models.

The development of agrometerological and agroclimatological models include climate crop and soil. The availability of water (which is an observed parameter through rainfall) and water requirement (expressed by PET which is an estimated parameter) are basic inputs in majority of these studies. The suitability of individual models for the estimation of PET depends not only on accurate predictive ability of the model but also on the availability of necessary input data. In addition, the following points shall be taken into considerations in model development and their application as precautions in agricultural meteorological studies.

1. Adequate human resource capacity has to be improved to develop and validate simulation models across the globe.

2. Multi-disciplinary research activities are essential for qualitative works in crop growth modeling.

3. Linkages shall be made and strengthened among research – education –extension to quickly disseminate the outcomes of models to farmers.

4. Majority of models use rainfall data at monthly interval, which is too long a period when compared to short duration of dry land crops (Sorghum, pearlmillet mature in 100 days). So, there is an urgent need to correct this anomaly by selecting proper methodologies.

5. It is important to differentiate between models of interest for research and for practical application. So, models useful to farmers with minimum input parameters shall find right place in the society.

6. It is always preferable to use one or more methods / models in conjunction to get more realistic conclusions useful to the farmers, who are the ultimate beneficiary alongwith industries.

7. The requirements in terms of details for local agricultural planning and operations are quite different from regional agricultural planning. Therefore, while selecting / using any model, it must be kept in mind the clear objective of study on one hand and verifying with ground truth (irrespective of the claims of the modal author) on the other hand.

8. The results/output of the models must be interpreted in an appropriate way by integrating soil, weather and crop. This interpretation is the key for success of agrometerological models. So, application of crop growth models shall form a part of studies of agrometeorologists on a regular and more continuous basis.

9. One should trust contemporary models particularly those concerned with yields. Wide variations may be found in the yield predicted by different models for specific crop in a defined environment. There is a need to develop test and improve the models with similar basis till they achieve comparable success for use by farmers, extension workers industry, etc. The reason is that the farmer needs them for decision-making, because It was found that the model can be used to identity new sites suitable for development of crop which finally results in generation of income to them.

10. It will be important to link the crop simulation models to local short and long-term weather forecasts. This will improve the yield predictions and provide policy makers with advanced yield information to help manage expected famines and other associated problems.

REFERENCES

Allen, L.H., Valle, P.R., Jones, J.W. and Jones, P.H. 1998. Soybean leaf water potential responses to carbon dioxide and drought. Agronomy Journal 90: 375-383.

Hammer, G.L., Sinclair, T.R., Boote, K.J., Wright, G.C., Meinke, H. and Bell, M.J. 1995. A peanut simulation model. 1. Model development and testing. Agron. J. 87: 1085-1093.

ICAR 2000. Crop Statistics Reports. Indian Council of Agricultural Research, New Delhi, India.

Sivakumar, M.V.K., Gommes, R. and Baier, W. 2000. Agrometeorology and sustainable agriculture. Agric. and For. Meteorol. 103: 11-26.

Van Keulen, H. and Wolf, J. (ed). 1986. Modelling of agricultural production: weather, soils and crops. Simulation Monographs. PUDOC, Wageningen, The Netherlands.

18

Agricultural Monitoring with GIS

Crop growth and yield are determined by a number of factors such as genetic potential of crop cultivar, soil, weather, cultivation practices (date of sowing, amount of irrigation and fertiliser) and biotic stresses. However, generally for a given area, year-to-year yield variability has been mostly modeled through weather as a predictor using either empirical or crop simulation approach. With the launch and continuous availability of multi-spectral (visible, near-infrared) sensors on polar orbiting earth observation satellites (Landsat, SPOT, IRS, etc) remote sensing (RS) data has become an important tool for yield modeling. RS data provide timely, accurate, synoptic and objective estimation of crop growing conditions or crop growth for developing yield models and issuing yield forecasts at a range of spatial scales. RS data have certain advantage over meteorological observations for yield modeling, such as dense observational coverage, direct viewing of the crop and ability to capture effect of non-meteorological factors. Recent developments in GIS technology allow capture, storage and retrieval and visualisation and modeling of geographically linked data. An integration of the three technologies, viz., crop simulation models, RS data and GIS can provide an excellent solution to monitoring and modeling of crop at a range of spatial scales.

LAI ESTIMATION USING RS-DATA

The leaf area index (LAI), defined, as the total one-sided leaf area per unit ground area, is one of the most important parameters characterising a canopy. Because LAI most directly quantifies the plant canopy structure, it is highly related to a variety of canopy processes, such as evapotranspiration, light interception, photosynthesis, respiration and leaf litterfall. RS-based LAI estimation would greatly aid the application of LAI as input to models of photosynthesis, crop growth and yield simulation models, evapotranspiration, estimation of net primary productivity and vegetation/ biosphere functioning models for large areas. A number of techniques for space borne remote sensing data have been developed/tested, ranging from regression models to canopy reflectance model inversions with varying successes, which include (1) statistical models that relate LAI to band radiance or develop LAI-vegetation index relation, (2) biophysical models like Price, and (3) inversion of canopy reflectance using numerical model or LUT based model. Myneni *et al.* developed a simple approach for estimating global LAI from atmospherically corrected NDVI using NOAA-AVHRR data. Oneor three-dimensional radiative transfer models were used to derive land coverspecific NDVI-LAI relations of the form

$$LAI = a \times \exp(b \times NDVI + c)$$

where, coefficients a and c are determined by vegetation type and soil.

Chen *et al.* have described relations using NOAA-AVHRR simple NIR/Red ratio (SR). These equations are vegetation type dependent and are being used to generate Canada wide 1 km LAI maps every 10/11 day. The equations are summarised in Table-1 and require a background SR that is season dependent as an additional input. In case of another high repetivity coarse resolution sensor, VEGETATION onboard SPOT satellite, use of SWIR channel is made to compute a new vegetation index, namely Reduced Simple Ratio (RSR). RSR reduces between vegetation and understory/background effects, thus making possible use of simplified equations for retrieval of LAI (Table-1).

Table 1. Equations for obtaining regional LAI products

Sensor	Vegetation Type	Model
NOAA-AVHRR	Coniferous forest	LAI = (SR – Bc)/1.153
	Deciduous forest	LAI = –4.1 × ln[(16–SR)/(16–Bd)]
	Mixed forest	LAI = –4.45 × ln[(14.5 – SR)/(14–Bm)]
	Other (crops, scrub etc.)	LAI = –1.6 × ln{14.5–SR)/13.5]
SPOT-VEGETATION		LAI = RSR/1.242
		LAI = –3.86 ln (1–RSR/9.5)
		LAI = –2.93 ln (1–RSR/9.3)
		LAI = RSR/1.3

Bc, Bd, Bm are background NDVI for coniferous, deciduous and mixed forests, respectively.

Using MODIS data, onboard TERRA, it is now possible to obtain operationally generated eight-day composite 'LAI product', at a spatial resolution of 1km, which incorporates model and look-up-table based LAI retrieval algorithms as a part of MODLAND. However, there is a need to validate this product, before it can be utilised in operational applications. Pandya *et al.* describe results of a study to develop small area LAI maps using IRS-LISS-III data using field sampling and regression approach and using the generated maps to validate MODIS LAI product. The atmospheric measurements of aerosol optical thickness and water vapour content were performed concurrently with the LAI measurements at the time of satellite acquisition and were used to convert digital numbers into the ground reflectance. These images were geo-referenced and the fields within the region of interest, where LAI measurements carried out were identified on images. Using NDVI of these fields, empirical models based on site-specific NDVI-LAI relation were developed (Figure 1) and used to generate LAI maps for each acquisition and study site. The LAI images were aggregated to 1km spatial resolution and compared with MODIS LAI product and results indicated significant positive correlation between LAI derived from LISS-III data and MODIS data albeit with a positive bias, in the MODIS product (Figure 2).

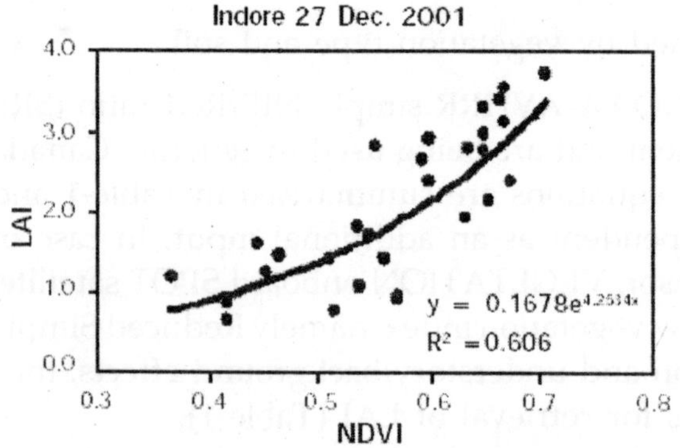

Figure 1: Relationship of ground measured LAI on wheat fields with IRS LISS-III derived NDVI

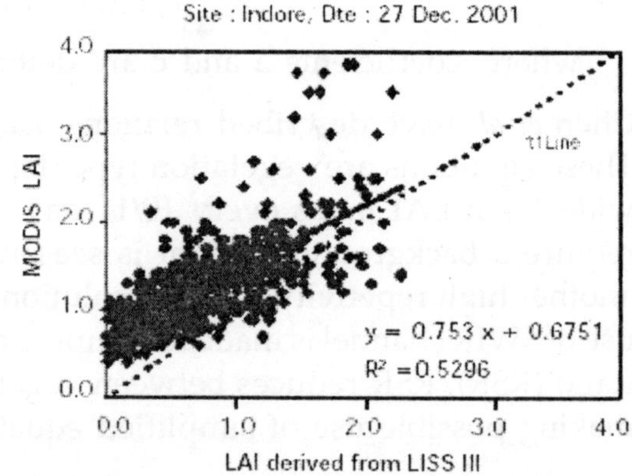

Figure 2: Comparison between LISS-III derived LAI and MODIS LAI

Rastogi *et al.* tested Price model on farmers fields during 1996-97 season in Karnal (Haryana, India) and 1997-98 in Delhi using IRS LISSIII data and estimated wheat attenuation coefficients. The root mean square error (RMSE) between RS estimates and ground measured LAI ranged between 0.78-0.87 when LAI was in the range of 1-4, while for higher LAI range (4-6), the RMSE varied from 1.25 to 1.5 in two sites. Such errors can severely reduce utility of a model using field-level LAI as input.

Crop Simulation Models

Crop simulations models are based on physical plant processes and simulate the effects of change in growing environment on plant growth and development on a daily basis. A crop simulation model is a simple representation of a crop and is explanatory in nature. The processes essentially modeled are phenology, photosynthesis and dry matter production, dry matter partitioning, in simulation models aimed at potential production. Those aiming at crop-specific behaviour include modules for phyllochron, branching pattern and potential flowers/grain filling sites. The response to water and nutrition limited environment is added by introducing models of soil water balance and uptake and transpiration by crop, and nitrogen transformations in soil, uptake and remobilisation within plant, respectively. Models of effects of weeds and pests are being developed and could be available in new generation of crop simulation models.

In dynamic crop simulation models, three categories of variables recognised are, state, rate and driving variables. The state variables are quantities like biomass, amount of nitrogen in soil, plant, soil water content, which can be measured at specific times. Driving variables, or forcing functions, characterise the effect of the environment on the system at its boundaries, and their values must be monitored continuously, e.g., meteorological variables. Each state variable is associated with rate variables that characterise their rate of change at a certain instant as a result of specific processes. These variables represent flow of material or biomass between state variables. Their value depends on the state and driving variables according to rules that are based on knowledge of the physical, chemical and biological processes that take place during crop growth. Under the International Benchmark Sites Network for Agrotechnology Transfer (IBSNAT) project a computer software package called the Decision Support System for Agrotechnology Transfer (DSSAT) was developed which integrates 11 crop simulation models (CERES cereal, CROPGRO legume and other models) with a standardised input and output and has been evaluated/used in a number of countries. Use of CERES-Wheat included in DSSAT for regional wheat yield prediction has been demonstrated recently in India.

GIS AND ITS USE FOR CROP MONITORING

GIS

Burrough and McDonnell has defined GIS as a powerful set of tools for collecting, storing, retrieving at will, transforming and displaying spatial data from the real world for a particular set of purposes. The three major components of GIS are (i) computer hardware, (ii) computer software and (ii) digital geographic data. The information stored within a GIS is of two distinct categories. The spatially referenced information that can be represented by points, lines, and polygons, that are referenced to a geographic coordinate system and is usually stored in either raster (grid-cell) or vector (arc-node) digital format. The second category of information stored in a GIS is attribute data or information describing the characteristics of the spatial feature.

Using RS & GIS for crop monitoring

The use of GIS along with RS data for crop monitoring is an established approach in all phases of the activity, namely preparatory, analysis and output. In the preparatory phase GIS is used for (a) stratification/zonation using one or more input layers (climate, soil, physiolgraphy, crop dominance etc.), or (b) preparing input data (weather, soil and collateral data) which is available in different formats to a common format. In the analysis phase use of GIS is mainly through operations on raster layers of NDVI or computing VI profiles within specified administrative boundaries. The final output phase also involves GIS for aggregation and display of outputs for defined regions (e.g., administrative regions) and creating map output products with required data integration through overlays. Wade *et al.* described efforts within National Agricultural Statistics Service (NASS) of U.S. Department of Agriculture (USDA) of using NOAA AVHRR NDVI for crop monitoring and assessment of damage due to flood and drought by providing analysts a set of map products. Combining satellite data in a GIS can enhance the AVHRR NDVI composite imagery by overlaying State and county boundaries. The use of raster-based (grid-cell) capabilities of ARC/INFO (GRID) for the generation of difference image helps compare a season with previous year or average of a number of years. Overlaying a crop mask helps in highlighting only effects on crops. Application of frost isolines is made to help analysts to locate average dates of the first frost for possible crop damage. Generation and overlay of contours of precipitation data generated using TIN function of ARC/INFO also is an aid to interpreting NDVI difference image.

Interfacing Crop Simulation Models to GIS

Crop simulation models, when run with input data from a specific field/ site, produce a point output. The scope of applicability of these simulation models can be extended to a broader scale by providing spatially varying inputs (soil, weather, crop management) and policy combining their capabilities with a Geographic Information System (GIS). The main purpose of interfacing models and GIS is to carry out spatial and temporal analysis simultaneously as region-scale crop behaviour has a spatial dimension and simulation models produce a temporal output. The GIS can help in spatially visualising the results as well as their interpretation by spatial analysis of model results. While GIS and modeling tools have existed for so long, the integration, including the conceptual framework is being given attention only recently. Hartkamp *et al.* have reviewed GIS and agronomic modeling and suggested that 'interface' and 'interfacing' be used as umbrella words for simultaneously using GIS and modeling tools, and 'linking', 'combining' and 'integrating' as suitable terminology for degree of interfacing. These correspond to loose, tight and embedded coupling, respectively, as used by Burrough and Tim. While there is a continuum between linking and combining, the terms are explained below:

1. *Linking:* Simple linkage strategies use GIS for spatially displaying model outputs. A simple approach is interpolation of model outputs. An advanced strategy is to use GIS functions (interpolation, overlay, slope, etc.) to produce a database containing inputs of the model and model outputs are also exported to the same database. Communication between GIS and model is through identifiers of grid cells or polygons in input and output files, which are transferred in ascii or biary format between GIS and model (Figure 3a). Such an approach is not able to utilise full potential of the system and suffers from limitation due to (a) dependence on formats of GIS and model, (b) incompatibility of operating environments and (c) not fully utilising the capabilities of GIS.

2. *Combining:* Combining also involves processing data in a GIS and displaying model results, however, the model is configured with GIS and data are exchanged automatically. This is done with facilities in GIS package of macro language, interface programmes, libraries of user callable routines (Figure 3b). This requires more complex programming and data management than simple linking. Example of combining is AEGIS (Agricultural and Environmental GIS) with ArcView.

3. *Integrating:* Integration implies incorporating one system into the other. Either a model is embedded in GIS or a GIS system is included in a modeling system. This allows automatic use of relational database and statistical packages (Figure 3c). This requires considerable expertise, effort and understanding of the two tools.

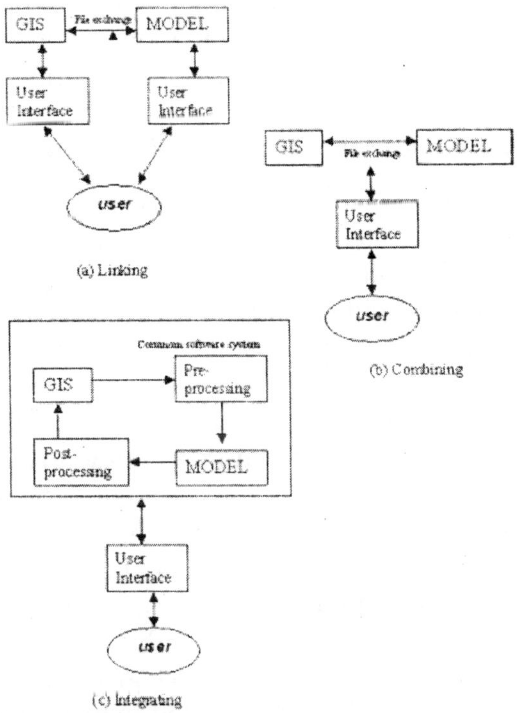

Figure 3: Organisational structure for (a) linking, (b) combining and (c) integrating GIS and crop models

Calixte *et al.* developed a regional agricultural decision support system, known as Agricultural and Environmental Geographic information System (AEGIS) that uses the DSSAT capabilities within ARC/INFO GIS for regional planning and productivity analysis. AEGIS allows the user to select various combinations of crop management practices over space and evaluate potential crop production. Engel *et al.* modified the AEGIS into AEGIS/WIN (AEGIS for Windows) written in Avenue, an object-oriented macro scripting language, which links the DSSAT (Version 3) with the geographical mapping tool ArcView-2. Thornton *et al.* developed spatial analysis software for the most recent release of the DSSAT, Version 3.1. This software standardised the links between crop models and GIS software and this allowed developers to make use of whatever GIS software is most suitable for a particular purpose, while ensuring that basic links to the DSSAT system and the crop models are the same. The spatial analysis software has two modules: (i) a geostatistical module to interpolate maps and produce probability surfaces from a network of data points, and (ii) a utility that calculates agronomic and economic output statistics from model simulations and maps the results as polygons. Another effort is development of a SPATIALEPIC linked to Arc/Info.

Demonstrated applications of CSM interfaced with GIS

Current wide range of applications of interfacing of GIS and modelling are reviewed by Hartkamp *et al.* and covers spatial yield calculation (regional and global), precision farming, climate change studies, and agroecological zonation, etc.

CGMS (Crop Growth Monitoring System) of MARS (Monitoring Agriculture with Remote Sensing)

This project of European Union uses WOFOST model and Arc/Info for operational yield forecasting of important crops. The Crop Growth Monitoring System (CGMS) of the MARS integrates crop growth modelling (WOFOST), relational batabase ORACLE and GIS (ARC/INFO) with system analytical part for yield forecasting. There are databases on soil, weather, crop, and yield statistics that cover the whole of EU. The system-analytical part consists of three modules: agrometeorological module, a crop growth module and a statistical module. The meteorological module takes care of the processing of daily meteorological data that are received in real time to a regular grid of 50×50 km for use as input by crop growth model or for assessment of 'alarm' conditions. The crop growth module consists of the dynamic simulation model WOFOST in which crop growth is calculated and crop indicators are generated for two production levels: potential and water-limited. In CGMS, WOFOST is run on a daily bases for each so-called 'simulation unit', i.e. a unique combination of weather, soil, and crop (mapping) units. In the statistical module, crop indicators (total above ground dry weight and dry weight storage organs) calculated with WOFOST are related to historical yield statistics through regression analysis in combination with a time-trend, for at least 15 years of simulated and historical data. The resulting regression equations per crop per region are used to make actual yield forecast. CGMS generates on a 10 day and monthly basis three types of output on current cropping season: (i) Maps of accumulated daily weather variables on 50x50 km grid to detect any abnormalities, e.g. drought, frost, (ii) Maps of agricultural quality indicators based on comparison of simulated crop indicators with their long-term means, (iii) Maps and tables of yield forecasts.

Precision Farming

Han *et al.* developed an interface between PC ARC/INFO GIS and SIMPOTATO simulation model to study potato yield and N leaching distribution for site-specific crop management (precision farming) in a 50 ha field. The GIS input layers, corresponding to important distributed input parameters for the model, were irrigated water/N layer, soil texture layers and initial soil N layers. For each unique sub-area stored in the GIS database, the interface programme extracts the attribute codes of that sub-area from the GIS database, converts the attribute codes to the input parameters of the SIMPOTATO and sends them to the model. After running the model, the interface programme retrieves the output data (potato yield and N leaching), converts them to the attribute codes and stores the output data in the GIS database.

Agro-ecological Zonation

Aggarwal used WTGROWS to simulate potential and water-limited wheat yields for 219 weather locations spread all over the country. The district boundaries (as polygons) and model input parameters of soil, weather stations and agro-ecological regions were stored in ARC/INFO GIS. The model outputs of potential and rainfed productivity were stored in GIS as polygon attribute data. Based on potential and rainfed productivity, the districts were classified into 10 iso-yield zones and represented as map using GIS.

Evaluating Agricultural land use options

Aggarwal *et al.* studied the agricultural land use option for the state of Haryana using symphonic use of expert knowledge, simulation 274 Crop Growth and Productivity Monitoring and Simulation modelling, GIS and optimisation techniques. The study area was divided into agro-ecological land units by overlaying maps of soil, soil organic carbon and climatic normal rainfall in raster GIS IDRISI. The original soil mapping units based on 19 soil properties were reclassified based on soil texture, level and extent of salinity and sodicity, slope and ground water depth. The organic carbon and normal rainfall maps were generated by inverse square interpolation of observed data points followed by segmentation. The CSM for specific crops have been linked to GIS layers of administrative boundaries, physiographic features, climate, soil and agroclimatic zones and GCM outputs to study effect of future climatic changes on crop potential/productivity.

LINKING CROP SIMULATION MODELS TO RS INPUTS & GIS

The use of remotely sensed information to improve crop model accuracy was proposed as early as two decades ago by Wiegand *et al.* and Richardson *et al.*. They suggested using spectrally derived LAI either as direct input to physiological crop model or as an independent check to model calculation for its re-initialisation. The main advantage of using remotely sensed information is that it provides a quantification of the actual state of crop for large area using less labour and material intensive methods than *in situ* sampling. While crop models provide a continuous estimate of growth over time, remote sensing provides a multispectral assessment of instantaneous crop condition with in a given area. The different ways to combine a crop model with remote sensing observations (radiometric or satellite data) were initially described by Maas and this classification scheme was revised by Delecolle *et al.* and by Moulin *et al.*. Five methods of remote sensing data integration into the models have been identified:

1. the direct use of a driving variable estimated from RS data in the model;
2. the updating of a state variable of the model (e.g., LAI) derived from RS ('forcing' strategy);
3. the re-initialisation of the model, i.e., the adjustment of an initial condition to obtain a simulation in agreement with the RS derived observations;
4. the re-calibration of the model, i.e., the adjustment of model parameters to obtain a simulation in agreement with the remotely-sensed derived observations, also called 're-parameterisation' strategy;
5. the corrective method, i.e., a relationship is developed between error in some intermediate variable as estimated from remotely sensed measurement and error in final yield. This relationship may be applied to a case in which final yield is not known.

Direct use of Driving variable

The driving variables of crop simulation models are weather inputs comprising daily observations of maximum and minimum temperature, solar radiation, relative humidity and wind speed as a minimal subset. In a recent review on this subject, Moulin *et al* cited inadequate availability of RS-derived parameters, due to cloud cover problem and intrinsic properties of sensors and platforms, as a major drawback, for adoption of this approach. However, this is a promising area, given the sparse distribution of weather observational network and recent progress in deriving some of these variables from sensors in space. Rainfall, solar radiation and intercepted/absorbed PAR have received maximum attention. Maas estimated the ratio of daily absorbed PAR (Q) to integrated daily PAR ® from radiometric NDVI and generated daily values of Q/R by linear interpolation between NDVI measurements for use as driving variable in a simplified maize growth model. The model showed an overestimation of 6.2% in above ground biomass at anthesis. METEOSAT based decadal (10-day) rainfall using cold cloud duration has been used as input to CERESMillet in Burkina Faso by Thornton *et al.* to forecast provincial millet yields halfway through crop duration to within 15% of their final values.

Forcing Strategy

The forcing strategy consists of updating at least one state variable in the model using remote sensing data. LAI has been the most commonly updated state variable. The concept of a simple crop simulation model and its modification for RS-derived LAI forcing is illustrated in Figure 4. Some examples of forcing spectrally derived LAI in crop simulation models are summarised in Table-2. The forcing could either be done only on day of RS observation or daily LAI profile is generated using some simple parametric model for use.

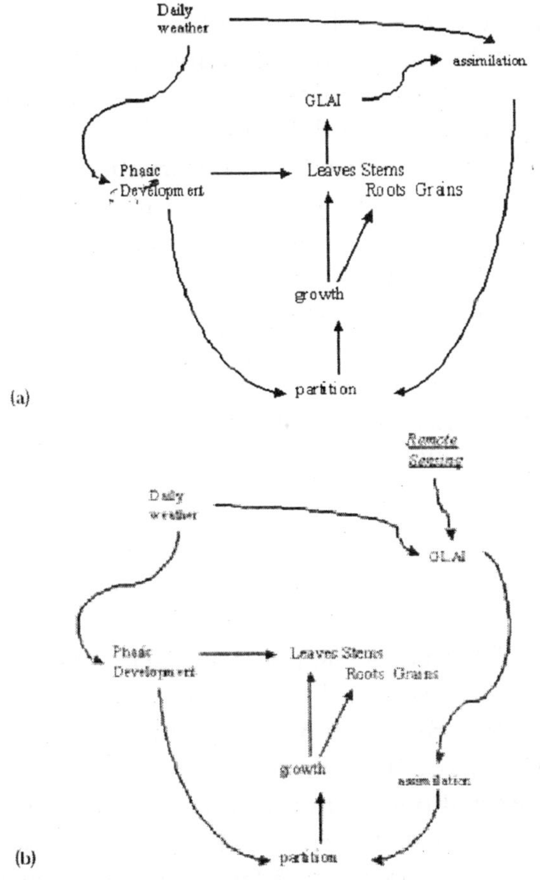

Figure 4: (a) Simple schematic of a crop simulation model. (b) Modified structure of crop simulation model with RS-based LAI forcing

Table 2: Selected case studies on use of RS-derived LAI for forcing crop simulation models

Crop	Model	LAI estimation & Interfacing	Evaluation of performance	Reference
Maize	–	Ground NDVI-LAI on obs. Dates	AGDM estimation improved	[1]
Wheat	AFRCWHEAT	SPOT/HRV LAI-WDVI relation: Daily interpolated LAI	Yield RMSE decreased	[2]
Wheat	SUCROS	LAI-WDVI	Biomass at harvest	[3]

[1] Maas, 1988a; [2] Delecolle and Guerif, 1988; [3] Bouman, 1995

Re-initialisation Strategy

The re-initialisation method takes advantage of the dependence of model performance on state variable initial condition. It involves adjustment of initial condition of state variable so as to minimise the difference between a derived state variable or the radiometric signal and its simulation. Maas in his simplified maize model adjusted the initial value of LAI (L0) at emergence based on the minimisation of an error function between remotely sensed LAI values and simulated LAI values during the course of simulation. Reinitialisation using one observation produced results similar to updating (forcing). However, the stability of model estimates obtained through reinitialisation increased as more observations were used. The observation at 51 days after emergence, which caused a 42% error using updating, resulted in less than a 3% error using re-initialisation. Maas demonstrated a similar study for sorghum using satellite observations. The simulation model was developed and verified using 10 fields in Central Texas in 1976 and the re-initialisation approach was validated for the 37 fields in South Texas using Landsat MSS data and agronomic observations. Without using the initialisation procedure, the average yield for the 37 fields was underestimated by 30%. Use of satellite derived green LAI data to initialise the same simulations resulted in a 2% overestimation of average yield.

Re-calibration/Re-parameterisation Strategy

In this approach it is assumed that model is formally adequate but requires re-calibration. This is achieved by minimising error between RS-derived state variable and its simulation by the model. This makes such an approach sensitive to errors in deriving state variables from RS data. In this case also, the state variable matched is LAI. However, depending on the model structure, which parameter to tune and number of observations used in analysis is critical. Maas demonstrated the re-calibration for maize model with remotely sensed GLAI observations. A multi-dimensional error function minimisation procedure was used which indicated more consistent estimates of LAI and biomass at anthesis as the number of parameters increased in multidimensional re-parameterisation. Delecolle *et al.* illustrated the use of re-calibration for rice crop using GRAMI model. Values of one to four parameters in the GRAMI model were re-calibrated to match the simulated LAI profile to observed LAI values. The results showed that improvement in simulated LAI profile by re-calibration depends largely on the number and timing of LAI observations. Clevers and Leeuwen used ground and airborne radiometric measurements over sugar beet fields to calibrate the SUCROS model. They derived LAI from measurements in optical and microwave wavebands. The adjusted parameters and initial conditions were sowing date, a growth rate, light use efficiency and maximum leaf area. The results showed that re-calibration with both optical and microwave observations estimated yield better than optical data alone. In the absence of optical remote sensing data, radar data yielded a significant improvement in yield estimation with the case of no remotely observed observations. Inoue *et al.* related paddy vegetation indices to the fraction of absorbed photosynthetically active radiation (fAPAR) as exponential equations with different parameters for the periods before and after heading. A real time recalibration module based on a simplex algorithm was developed and proved effective in linking remotely sensed fAPAR with a simple rice growth model.

Re-parameterisation using Coupled Crop Simulation Models and Canopyradiation Models

The re-initialisation and re-parameterisation of crop models can also make direct use of radiometric information instead of deriving canopy parameters from them. In this strategy, coupling a radiative transfer reflectance model to the crop production model reproduces the temporal behaviour of canopy surface reflectance, which can be compared with canopy reflectance observed from satellite. Adjusting initial conditions or model V.K. Dadhwal 279 parameters carries out the minimisation of differences between the simulated and observed reflectance values. Such an approach has been used by Clevers *et al.* and Guerif and Duke for sugarbeet. The LAI simulated by SUCROS on dates of spectral observations was passed on to PROSPECT-SAIL and SAIL model, respectively, and parameters of SUCROS model adjusted to minimise differences between observed WDVI and simulated WDVI. The reparameterisation of SUCROS reduced yield prediction errors. This approach was extended to microwave RS by Bouman *et al.* who simulated radar backscatter of agricultural crops (sugar beet, potato, winter wheat) by integrating LAI and leaf moisture from SUCROS with top soil moisture content by soil water balance model (SAHEL) and radar backscatter model (CLOUD). Since canopy-radiation models such as SAIL have parameters in addition to LAI, their uncertainty could affect the results from this approach. Moulin and Guerif concluded that the error in canopy reflectance estimates as a result of omitting data on canopy leaf angle and soil reflectance, two parameters in SAIL model, is so large that direct use of simulated canopy reflectance in simulation models for yield prediction is severely affected. However, the use of vegetation indices drastically reduced the errors linked to crop structure (NDVI) and to soil reflectance (TSAVI).

Corrective approach

Sehgal *et al.* used this strategy for generating the wheat yield maps for farmers' fields during rabi 1998-99 in Alipur block (Delhi). The RS inputs as estimated LAI were linked to wheat simulation model WTGROWS for yield mapping and results were validated with yield observations on farmers' fields. Biometric relation of grain yield and leaf area index (LAI) is derived from simulation model by running model for a combination of input resources, management practices and soil types occurring in the area. Then this biometric relationship is applied to all the crop fields of the study area for which the LAI is computed from remote sensing data. The WTGROWS simulated grain yield for the combination of inputs showed yields varying between 1.1 and 4.9 t ha^{-1}. The corresponding range of simulated LAI on 27 Jan 1999 was 0.6 to 4.2. The regression equation fitted between simulated LAI on 27th Julian day (i.e. 27-January-99) and simulated grain yield showed saturating logarithmic nature with a R^2 value of 0.81. The relationship is given below:

$$\text{Yield (kg/ha)} = 1571.2 * \ln(\text{LAI}) + 2033.6 \quad (2)$$

This empirical biometric relation was applied to the LAI map of the wheat pixels and grain yield map for farmer's fields of Alipur block, Delhi, was generated. The predicted yields ranged from 2.1 to 4.8 tha^{-1}. The comparison of predicted grain yield and observed yield for the 22 farmers' fields showed high correlation coefficient of 0.8 and a root mean square error (RMSE) of 597 kg ha^{-1} which was 17 per cent of the observed mean yield (Figure 5).

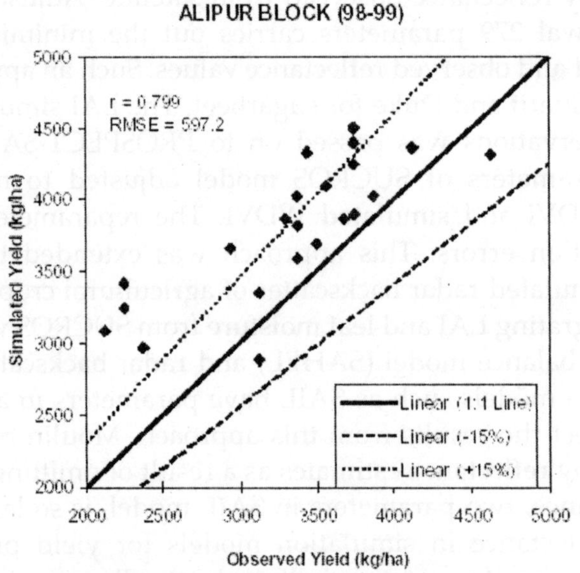

Figure 5: Comparison of predicted grain yield

Development of a RS-based CGMS for wheat in India

Sehgal *et al.* reported the development of a prototype Crop Growth Monitoring System (CGMS) for wheat using WTGROWS simulation model on a 5′×5′ grid in GIS environment for generating daily crop growth maps and predicting district-wise grain yield. The inputs used were RS based wheat distribution map, daily weather surfaces, soil properties map and crop management input databases in a GIS environment and analysis for wheat season of 1996-97 was carried out. The comparison of simulated grain yields aggregated at district level and estimates by the State Department of Agriculture is shown in Figure 6. In general, the model simulated yields were higher than observed. This could be due to a number of yield reducing factors such as pest, weed, soil constraints, which operate in field but are not considered by the model. The model predicted yields were within ±10% of reported yields in 12 out 16 districts. The RMSE of 335.4 kg ha^{-1}, which is less than 10 per cent of the State mean yield, was obtained. Only in two districts, Mahendragarh and Bhiwani, the simulated district yields were lower than observed yields while for Kaithal, Karnal, Ambala and Yamunanagar, the simulated yields were higher than observed yields by more than 10 per cent.

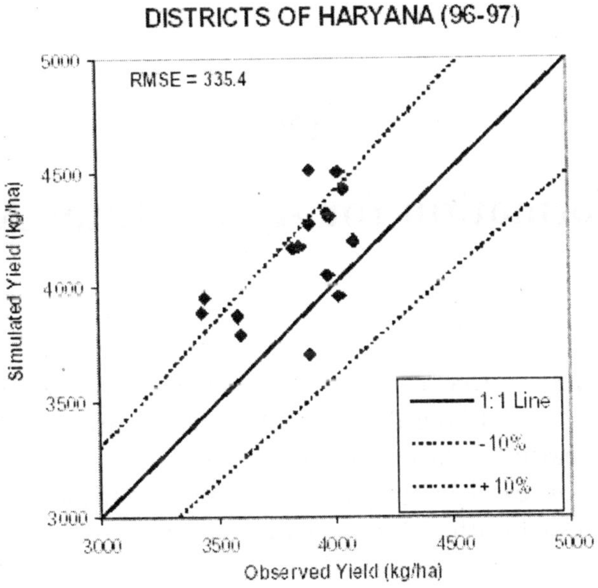

Figure 6: Comparison of simulated district wheat yields

Sehgal *et al.* demonstrated a technique for estimating date of sowing (DOS) using RS-derived spectral-temporal crop growth profiles and CGMS simulation capability and evaluated the capability of CGMS for spatial yield mapping and district level yield prediction for Haryana State during 2000-01 crop season. The technique for estimating district-wise DOS matched the RS-derived date of peak NDVI (from multi-date WiFS sensor aboard IRS-1D satellite) to date of peak LAI simulated in CGMS for a range of plausible dates of sowing.

REFERENCES

Bacsi, Z., Thorton, P.K. and Dent, J.B. 1991. "Impacts of future climate change on Hungarian crop production: An application of crop growth simulation models". *Agric. Sys.*, **37**: 435-450.

Engel, T., Hoogenboom, G., Jones, J.W. and Wilkens, P.W. 1997. "AEGIS/WIN - a programme for the application of crop simulation models across geographic areas". *Agronomy J.*, **89**: 919-928.

Qiu, J., Gao, W. and Lesht, B.M. 1998. "Inverting optical reflectance to estimate surface properties of vegetation canopies". *Int. J. Remote Sens.*, **19**: 641-656.

Tim, U.S. 1996. "Coupling vadose zone models with GIS: Emerging trends and potential bottlenecks". *J. Environ. Qual.*, **25**: 535-544.

Wiegand, C.L., Richardsons, A.J. and Kanemasu, E.T. 1979. "Leaf area index estimates for wheat from LANDSAT and their implications for evapotranspiration and crop modeling". *Agronomy J.*, **71**: 336-342.

19

Satellite Information for Agricultural Applications

Satellites offer a unique source of information for many agricultural applications. Among their attributes are: (1) regular, repetitive observation patterns; (2) large scale synoptic view of the earth surface from space; and (3) availability of a permanent information archive for establishing baseline data. Although a large number of atmospheric parameters can affect the agricultural production in short and long time scale.

1. Clouds
2. Rainfall
3. Soil Moisture
4. Solar Radiation
5. Surface Temperature
6. Temperature and Humidity Profiles.

Meteorological satellites play an important role in retrieval of the above parameters at large spatial scales.

CLOUDS

Interpretation of Cloud in Visible/Infrared Imageries

The cloud configuration seen in satellite imagery represents a visible manifestation of all types of atmospheric processes. The complete interpretation of cloud structure must make use of both satellite imagery and other available observational data. One should remember that the satellite sensors view only tops of clouds while in surface observations their bases are seen. Basically following six characteristic features of satellite pictures are helpful in extracting information for weather forecasting:

1. *Brightness:* The brightness depends strongly upon the albedo of the underlying surface in visible pictures. Highly reflecting surfaces like cumulonimbus (Cb) tops and snow appear pure white whereas the sea looks black. Other clouds and land appear in varying degrees of gray. In IR images warm land surfaces appear very dark while cold ones are white (e.g. Cb tops, thick cirrus, snow etc.). Lower level clouds and thin cirrus appear gray.
2. *Pattern:* Cloud elements are seen to be organised into identifiable patterns like lines, bands, waves etc.
3. *Structure:* In a VIS image, shadows of taller clouds fall on lower surfaces. Shadows and highlights thus give an idea of the cloud structure. In IR imagery this information is provided by the cloud top temperature (CTT) more directly.
4. *Texture:* The cloud surfaces when viewed by the satellite vary in degree of apparent smoothness. Some clouds appear smooth while some may look ragged.
5. *Shape:* Clouds assume a variety of shapes—rounded, straight, serrated, scalloped, diffused or curved.
6. *Size:* The size of a pattern or the size of individual elements in a pattern are useful indicators of the scale of weather systems.

While interpreting satellite pictures continuity in time has to be maintained. The pictures should not be viewed in isolation but must be interpreted with reference to past weather and earlier imageries. It is necessary to keep in mind the time of the day, season and local peculiarities while interpreting satellite imageries. In VIS pictures illumination depends on the position of the sun which will vary the brightness of clouds. Similarly in different seasons (e.g. summer and winter) the image disc of the northern and southern hemisphere will have different brightness. Local features like mountains and valleys introduce their own effects. In IR imagery elevated land like Tibetan Plateau will be cold at night and would appear very bright whereas tropical oceans would maintain about the same gray shade throughout the diurnal cycle.

Interpretation of Visible Imageries

Clouds have higher albedo than land (apart from snow cover) and appear white or light grey in a VIS imagery. Their brightness depends on their physical properties. Clouds with high albedo have large depth, high cloud water (ice) content, small cloud-droplet size whereas clouds with low albedo have shallow depth, low cloud water (ice) content, large average cloud-droplet size. The water content and depth of the cloud are the most important. Typical Albedo values are given in Table 1.

Table 1. Albedo values of different surfaces and clouds

Earth surface	(%)	Clouds	(%)
Oceans, Lakes	8	Shallow broken clouds	30
		Cu, Ci, Cs, Cc	35
Land Surfaces	14-18	St	40
Sand, Desert	27	Thick clouds (Cs)	74
Ice and Snow		Ac, As, Sc	68
Sea ice	35	Cu	75
Old snow	59	Ns	85
Fresh snow	80	Cb	90

VIS imagery is useful for distinguishing between sea, land and clouds (Figure 1). Seas and lakes have low albedo and hence appear dark. Land appears brighter than sea but darker than clouds. Albedo of land varies with the type of surface. Deserts appear very bright in contrast to the darkness of forests and vegetated areas. When the sun shines obliquely onto clouds the shadow thrown by an upper cloud layer onto a lower layer reveals the vertical structure of the cloud in VIS imagery.

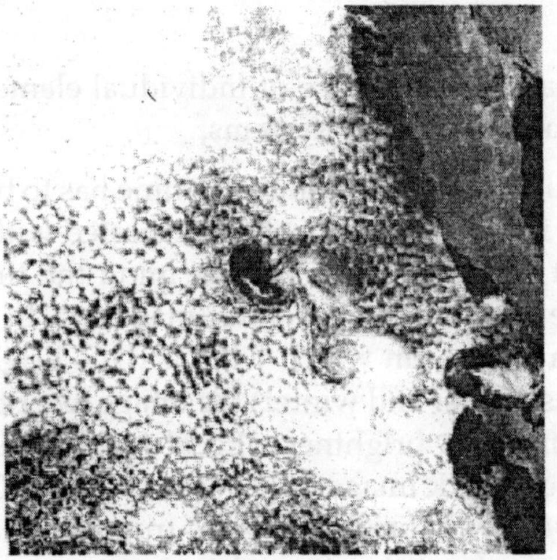

Figure 1: Karman cloud vortices in Pacific Ocean as they are seen in visible images.

The texture of the cloud in VIS imagery can help in its identification, (for example) its cellular pattern can distinguish stratocumulus clouds (Sc) from stratus (St). No VIS imagery can be obtained at night. To distinguish clouds from snow covered ground a knowledge of the surface topography is essential. Thin clouds have low albedo and do not show up very brightly in VIS imagery so that the cloud cover over dark surfaces may be underestimated. In the same manner thin cloud over a high albedo desert surface may look misleadingly bright and thick. Mesoscale cumulus (Cu) clouds which are smaller than the resolution of the satellite will be depicted in rather lighter grey shades in VIS imagery quite unlike the normal view of convective clouds.

Interpretation of IR Imagery

IR imagery indicates the temperature of the radiating surfaces. In black and white image warm areas are shown in dark tones and cold areas in light tones. Clouds will generally appear whiter than the earth surface because of their lower temperature. In this respect IR and VIS images have some similarity. Because cloud top temperature decreases with height IR images show good contrast between clouds at different levels. This is not possible in VIS imagery.

Coast lines show up clearly in IR images whenever there is strong contrast between land and sea surface temperatures. During the day the land may appear darker (warmer) than the sea but at night may appear lighter (cooler). At times when land and sea temperatures are almost same, it becomes impossible to detect coastlines in IR imagery. The most marked contrast between land and sea is normally found in Summer and Winter and is least in Spring and Autumn. Thin Ci which is often transparent in the VIS can show up clearly in IR especially when it lies over much warmer surface. IR imagery is inferior to VIS in providing information about cloud texture as it is based upon emitted and not scattered radiation. Low clouds and fog can rarely be observed in IR at night because they have almost the same temperature as the underlying surface. But during day, such clouds are easily detected in.

Interpretation of WV Imagery

Water vapor imagery is derived from radiation at wavelengths around 6-7 μm. Though this is not an atmospheric window, it is part of the spectrum where water vapor is the dominant absorbing gas. It has a strong absorption band centered on 6.7 μm. In regions of strong absorption, most of the radiation reaching the satellite originates high in the atmosphere. The stronger the absorption, the higher is the originating level of the emission that ultimately reaches the satellite. As the Relative Humidity (RH) decreases the main contribution of the radiance received by the satellite comes from lower in the troposphere. WV imagery is usually displayed with the emitted radiation converted to temperature like IR imagery. Regions of high upper tropospheric humidity appear cold (bright) and regions of low humidity appear warm (dark) i.e. when the upper troposphere is dry, the radiation reaching the satellite originates from farther down in the atmosphere where it is warmer and appears darker on the image. In a normally moist atmosphere most of the WV radiation received by the satellite originates in the 300-600 hPa layer. But when the air is dry some radiation may come from layers as low as 800 hPa. Due to the general poleward decrease of water vapor content, the height of the contributing layer gets lower and lower towards the poles.

Since clouds do emit radiation in this wave band, high clouds may be seen in this type of imagery. Thick high clouds as Cb anvils stand out prominently in both WV and IR imageries. Broad-scale flow patterns are particularly striking in WV imagery. This is because WV acts as passive tracer of atmospheric motions. WV imagery is therefore useful for displaying the mid tropospheric flow, upper tropospheric cyclones are defined clearly by moist spirals or comma-shaped patterns. Subsidence areas appear dark. Jet streams are delineated by sharp gradients in moisture with dry air on the poleward side. Even when a WV image indicates a very dry upper troposphere, there may well be moist air near the surface. Moist air or cloud in the lower half of the troposphere is not depicted well in WV imagery.

RAINFALL

Rainfall is not only an important parameter for agro-meteorology, it is also a vital component of earth's hydrological cycle. Oceans receive heat from the sun, and the evaporation takes place. When the same evaporation travels vertically, it gets condensed to form clouds and eventually precipitation. During this process, enormous amount of energy is released which is called the latent heat of condensation. This is one of the major sources of energy that drives the circulation in tropical atmosphere. Hence, the knowledge of rainfall and its distribution at current time is also important for its future prediction.

There are several techniques to derive rainfall from satellite observations. The earliest developed methods that are useful even today are based on the visible/infrared observations from satellites. Techniques based on visible sensors rely on the identification of cloud types. Each cloud type is assumed to have different rain intensity, and then the rain is derived based upon the extent of each cloud type. In infrared-based methods, the most common approach is to find cold clouds (say, colder than 250°K) within an overcast area. The raining potential of the clouds is proportional to the fractional area covered by cold clouds, and thus the rainfall derived. More complex techniques use both visible and infrared observations to create a bi-spectral histogram of the cloud images. Bi-spectral histogram method is a simple technique in which the clouds can be classified based on the combination of cloud signatures in visible and infrared frequencies. Then the rainfall is derived by estimating the extent of each type of cloud and multiplying it by the a-priori rain potential of respective classes. However, all the rain-retrieval techniques based on visible/IR observations are basically "inferential" in nature, because these sensors can sense the clouds (that too, the top surfaces of the clouds) but not the actual rain, that occurs at several layers below the clouds. Visible/IR techniques make a "guess" about the rainfall based on the cloud features. Due to this shortcoming, the estimates of rainfall based on visible/IR technique are not very accurate on instantaneous time scale. However, long time averages (e.g. daily, weekly, and monthly) of rainfall are better and usable for practical purposes.

On the other hand, rainfall estimation techniques based on microwave frequencies (0.1 cm to 100 cm wave length) are more direct in nature. Due to their large wavelengths, these frequencies can easily penetrate clouds. However, these frequencies interact effectively with rainfall. Let us consider the case of passive microwave methods. In these techniques, the microwave instrument onboard satellite does not have any source of microwave illumination. It has just a receiver that can gather the microwave emission coming from earth, ocean or atmosphere. Due to small emissivity in microwave region, ocean surface emits small amount of microwave radiation. When the rainfall occurs over a layer in the atmosphere, two different processes take place. The atmospheric rain layer itself emits microwave radiation and thus the radiation received at satellite is greater than the radiation received in no-rain situation. This process is predominant at lower frequencies (e.g. 19 GHz, or about 1.5 cm wavelength). On the other hand, the rain drops, ice and snow particles, scatter the microwave radiation (particularly at higher frequencies, e.g. 85 GHz, or about 0.3 cm wavelength) that is coming up from the ground. In this case the radiation received at satellite will be smaller than that in norain situation. In both the cases, the change in the microwave radiation (measured in terms of brightness temperature) can be related to the intensity of rainfall. Various algorithms have been developed in the past that use either low frequency or high frequency, or a combination of both. It is to be noted that for emission based algorithms (using lower frequencies of microwave e.g. 19 GHz), it is important that the emission from the background should be uniform and also as little as possible, so that the emission from rainfall can be detected clearly. So emission based algorithms are effective only over the ocean surfaces, while the scattering based algorithms using higher frequencies of microwave, can be used over the ocean as well as over the land. Special Sensor Microwave Imager (SSM/I), and TRMM Microwave Imager (TMI) are good examples of passive microwave sensors that are quite effective in the determination of global rainfall.

SOIL MOISTURE

Recent advances in remote sensing have shown that soil moisture can be measured by a variety of techniques. However, only microwave technology has demonstrated a quantitative ability to measure soil moisture under a variety of topographic and vegetation cover conditions so that it could be extended to routine measurements from a satellite system. Both active and passive microwave techniques have been applied by researchers for the estimation of soil moisture. Two material properties provide clues about composition and surface state by the manner in which these attributes interact with the microwave radiation. One property is the dielectric constant (its symbol is the small Greek letter, *k*), which is the ratio of the capacitance of a material to that of a vacuum. It is a dimensionless number that is set at 1.00. This electrical property describes a material's capability (capacity) to hold a charge, which also measures its ability to polarise when subjected to an electric field. Microwave radiation penetrates deeper into materials with low dielectric constants and reflect more efficiently from those with high constants. Values for *k* range from 3 to 16 for most dry rocks and soils, and up to 80 for water with impurities. Moist soils have values typically between 30 and 60. Thus, variation in emitted microwave radiances (in case of passive microwave observations e.g. by radiometers) or reflected-pulse intensities (in case of active microwave observations e.g. by radar) may indicate differences in soil moisture, other factors being constant. Dry soil has a high emittance; water surfaces have low emittance in microwave region. If one adds water to the soil, the emittance falls and becomes polarised. With the knowledge of "normal" emittance at a particular location (which depends upon soil type and vegetation), microwave observations can be used to detect changes in emittance and therefore of soil moisture. Since the soil moisture is changed by precipitation, these emittance changes between two satellite passes can serve as a proxy of precipitation, known as Antecedent Precipitation Index (API).

The second material property is roughness that can be used to define the *soil texture*. Materials differ from one another in their natural or cultivated state of surface roughness. Roughness, in this sense, refers to minute irregularities that relate either to textures of the surfaces or of objects on them (such as, closely-spaced vegetation that may have a variety of shapes). Examples include the textural character of pitted materials, granular soils, gravel, grass blades, and other covering objects whose surfaces have dimensional variability on the order of millimeters to centimeters. The height of an irregularity, together with radar wavelength and grazing angle at the point of contact, determines the behaviour of a surface as smooth (specular reflector), intermediate, or rough (diffuse reflector). A surface with an irregularity height averaging 0.5 cm will reflect Ka band (l = 0.85 cm), X band (l = 3 cm), and L band (l = 25 cm) radar waves as if it were a smooth, intermediate, and rough surface, respectively. Other average heights produce different responses, from combinations of "all smooth" to "all rough" for the several bands used. This situation means radar, broadcasting three bands simultaneously in a quasimulti-spectral mode, can produce color composites.

SOLAR RADIATION

Incoming solar radiation is the primary source of energy for plant photosynthesis. Solar radiation also plays a key role in evapotranspiration. Visible observations from satellites provide an excellent source of information about the amount of solar radiation reaching the plant canopy. A measurement of solar energy reflected to space from earth-atmosphere system immediately specifies the maximum amount of solar energy that can be absorbed at the surface. Incoming solar radiation can be known by adjusting the amount absorbed. Hence, for the computation of downwelling solar radiation, the albedo of the surface must be known. This is especially important over the regions of high reflectivity such as snow and desert. Tarpley used a statistical regression technique to obtain surface fluxes over the land from Visible channel observations from geostationary satellites. In this model, cloud amount is estimated for a given location from satellite visible data. Three separate regression equations are then used to estimate solar radiation for three categories of clouds. This method provides an accuracy of 10% for clear sky, 30% for partly cloudy and 50% for overcast conditions. Other algorithms like those by Moser and Raschke, and Pinker and Ewing used physical approaches, and treated the interaction of incoming and reflected solar radiation with the atmosphere and land surfaces in physical manner. The transmittance of solar radiation in these approaches is solved by the use of radiative transfer equations that take into account the concentration profile of different atmospheric components. These physical schemes also take into account the cloudiness and atmospheric water vapor. These methods provide relatively higher accuracy. However, statistical techniques have remained the choice for operational use. These methods require coincident satellite and ground (pyranometer) observations to develop the coefficients in the regression equations. These methods produce daily total insolation, based on hourly estimates made from geostationary satellite data between 0800 and 1600 LST, with interpolation used toward both sunrise and sunset and for any other missing hourly values.

SURFACE TEMPERATURE

Air temperature is significantly related to crop development and conditions. Operational crop and soil moisture models require daily minimum and maximum shelter temperature and dew point temperature. Canopy (or skin) temperature may be more directly related to growth and evapotranspiration than the shelter temperature. The difference between the two is a measure of crop stress. The ability to observe canopy temperature directly is an advantage of satellite observations. However, it is to be noted that the satellite derived "skin temperature" and "crop canopy temperature" are equivalent only when a satellite field of view (FOV) is filled with vegetation. If FOV constitutes a mixture of bare soil, water bodies, etc. the relation between the two becomes complex. Satellite observations in the thermal IR window (10-12 μm) are used to obtain estimates of canopy or skin temperature. Price found that surface temperatures over vegetated land can be estimated with an accuracy of 2-3°C using AVHRR split window technique. The errors can largely be attributed to imperfect knowledge about atmospheric water vapor content and wavelength dependent surface emissivity. Clouds pose a serious problem in the estimation of surface temperature by infrared techniques. Satellite surface temperature estimates have been used to delineate areas of freezing for frost warning and for monitoring freezing events that can affect food production. In U.S., the infrared observations from geostationary satellite GOES are used for freeze forecasting by following the diurnal progression of the freeze line. Accurate freeze forecasts permit farmers to protect crops only when there is a significant freeze threat.

Shelter Temperature

Shelter temperature, its minimum/maximum values along with canopy temperature are important factors of consideration for assessment of crop development and crop stress.

Shelter temperature is more directly related to the air temperature than the surface temperature. Observations from atmospheric sounders like TOVS (TIROS-N Operational Vertical Sounder) are used in methods for estimating shelter temperature. A simple linear regression approach is generally used to derive shelter temperature from satellite soundings. Collocated and coincident sets of satellite soundings and shelter temperature observations are used to develop regression coefficients. This method provides the shelter temperature with an accuracy of about 2o C. In winter time, a temperature inversion generally persists over the cold ground. This results in a bias of 1-2°C (satellite estimates being warmer) in the nighttime estimates of shelter temperature during winter.

TEMPERATURE AND HUMIDITY PROFILES (SOUNDING)

Temperature and moisture structures of earth's atmosphere are some of the most significant factors that influence the weather and climate patterns on the local as well as on global scale. The vertical structure of temperature and humidity is retrieved by satellite "sounders" that operate in infrared and microwave frequencies. Sounders use the principle of gaseous absorption for the retrieval of temperature and humidity profiles. For the retrieval of temperature, absorption spectra of some uniformly mixed gas such as CO_2, or, O_2 is used (Fig. 3). These gases absorb earth's upwelling radiation. However, at some wavelengths (say, l1) their efficiency of absorption is very strong, while at some neighboring wavelengths (say, l2), it is very weak.

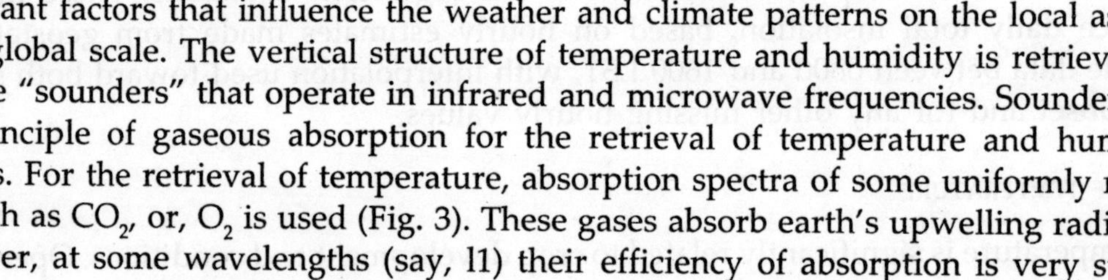

Figure 3: (a) Curve showing the absorption of infrared radiation by different gases in the atmosphere. (b) Numbers shown in this curve (1,2,3,4...7) denote the central wavelengths of CO_2 absorption band used for sounding of atmospheric temperature profiles

Now, any radiation of wavelength l1 coming from lower layers of atmosphere has very little chance of reaching up to satellite, because it is getting strongly absorbed by the given gas. So, at this wavelength, most of the radiation will be coming from the upper layers of the atmosphere. Similarly, the radiation emitted from the ground at wavelength l2 will reach the satellite without much interruption, because the atmospheric gases absorb this radiation very weakly. It means that the radiation at wavelength l2 contains information about the atmospheric layers near the surface. Similarly, radiation at other wavelengths lying between l1 and l2 is sensitive to different atmospheric layers in vertical.

The principle of the sounding of humidity profiles is similar to that of temperature. However, water vapor is not a uniformly mixed gas, and also it changes phase (i.e. ice, water, snow, or vapor) very frequently within the atmosphere. Satellite water vapor sounders use the water vapor absorption frequencies (~ 6 lm in IR sounder, and ~ 183 GHz in microwave sounders). However, the upwelling radiation at these wavelengths/ frequencies not only depends upon the water vapor amount in different atmospheric layers, but also on the temperature of those layers. In this case a-priori information about the temperature structure of the atmosphere is crucial for the retrieval of humidity profiles. An example of infrared sounders is TIROS Operational Vertical Sounder (TOVS), onboard NOAA series of satellites. This instrument is designed to profile temperature and water vapor. The TOVS is actually a three instrument complex: the High Resolution IR Sounder (HIRS-2), with 20 channels; the Stratospheric Sounding Unit (SSU), with three channels near 15 μm, and the Microwave Sounding Unit (MSU), a passive scanning microwave spectrometer with four channels in the 5.5 μm interval. Advance Microwave Sounding Units (AMSU-A, and AMSU-B onboard recent series of NOAA) are the examples of microwave sounders that are designed to sense temperature and humidity profiles respectively.

OBSERVATIONS OF THE EARTH'S SURFACE

The brightness of the earth's surface in VIS imageries depends on the time of the day, season of the year, the geographical location etc. It is easy to distinguish dry land from a water surface and to identify shorelines, rivers, lakes, islands etc. Dry land, depending on the type of relief, assumes various grey shades in VIS imageries. The tone of the underlying surface in IR imageries depends on temperature and so its appearance is affected on the latitudinal, diurnal and seasonal variations. During day light hours and in summer dry land surfaces appear darker than the water surface but at night it has a lighter tone than a water surface. IR imageries reveal only large irregularities of relief of vegetation pattern which are associated with marked temperature gradients whereas VIS pictures can reveal even small terrain variations.

Deserts and Vegetated Areas

Dense vegetated areas appear relatively dark in VIS imageries. Mountains with thick forests can be easily identified when sparsely vegetated plains surround it. Deserts with very sparse vegetation combined with red and yellow soils and rocks make the earth's surface highly reflective. These areas appear brightest in satellite pictures. Surface temperature differences between deserts and vegetated regions are often modified diurnally by water vapor absorption in moist low levels of the atmosphere. Hence the radiation emitted from the surface reaches the satellite undiminished only when the atmosphere is dry. When it is moist the radiation from the high cold levels only reach and hence the IR imagery is not much influenced by the surface characteristics.

Snow

Knowledge of surface geography together with an appreciation of the climatological variation of snow cover through the seasons is a basic prerequisite for the analysis of satellite imagery in respect of snow. The tone of snow on VIS imagery varies from bright white to light grey depending on the nature of the relief, vegetation, the age of snow and the illumination. In areas of relatively flat terrain without trees snow surfaces appear uniformly white in tone. In areas covered by extensive forests a snow-covered surface appears patchy, brighter patches correspond to areas where there are no trees and dark patches to forest areas. In IR imagery, snow covered terrain often appears whiter than its surroundings. At night the snow surface cools more rapidly than its surroundings and during the day it warms gradually. Light snow on the tops of high mountain ranges is less detectable in IR than in VIS data as the temperature difference between cold land and snow is small while the difference in albedo between snow and land is large.

ATMOSPHERIC POLLUTANTS

Dust and Haze

Dust is characterised by a dull, hazy and filmy appearance similar to thin cirrostratus. Dust can extend to adjacent water bodies and sometimes obscure the coastline. The edges of dust areas are ill defined. Blowing dust and sand can be observed in satellite imageries when the reflectivity of the suspended particles differs greatly from the reflectivity of the underlying surface. Haze particles scatter quite effectively at the blue end of the visible wavelength (channel 1) and hence areas of haze show up better in images of this channel. Particles of dust or sand carried by the wind form a cloud that causes land marks to appear blurred or to disappear in VIS images. To be detected in IR imagery a dust cloud must be very deep. There are two reasons for the same. Firstly the dust must be composed of large enough particles (over 15 m) to obscure radiation emitted from the ground at least partially. Secondly it must be deep enough to have low temperature otherwise it will not be able to distinguish it from the surface.

Forest Fires

The usefulness of weather satellites are not limited to meteorological observations alone. Their camera systems, sensors and their global coverage enable them to perform a variety of non-meteorological tasks as well. These are detection of forest fires, the tracking of locust clouds, observing volcanoes etc. The sensors of weather satellite enable detection of forest fires characterised by a dull, hazy appearance as in the case of dust. Detection becomes difficult in VIS imagery if the smoke is not prominent. Under such condition it might be possible to spot the fire in IR imagery by sensing the thermal radiation given off. But the detection depends on the intensity of forest fire, its extent, the resolution of the satellite, presence or absence of clouds etc. If a thick cloud system covers the area of forest fire or if clouds have been generated by hot air rising above the forest fire zone it may not be possible to detect forest fires efficiently.

REFERENCES

Ananthkrishnan, R. and Soman, M.K. 1988. "The onset of Southwest monsoon over Kerala: 99 (1901-1980)". *Journal of Climate,* 8, 283-296.

Arkin, P.A. and Ardanuy, P.E. 1989. "Estimating climatic scale precipitation from space: a review". *Journal of Climate,* 2, 1229-1238.

Bader, M.J., Forbes, G.S., Grant, J.R., Lilley, R.B.E. and Waters, A.J. "Images in weather forecasting, a practical guide for interpreting satellite and radar imagery". *Syndicate of Cambridge Uni.,* The Pitts bldg, Cambridge, CB21RP, New York, NY1001-4211, USA.

Das, P.K. 1968. "The Monsoon". Published by the Director, *Nat. Book Trust,* New Delhi.

Davis, P.A. and Tarpley, J.D. 1983. "Estimation of shelter temperature from operational satellite sounder data". *Journal of Climate and Applied Meteorology,* 22, 369-376.

Fein, J.S. and Stephens, P.L. 1986. "Monsoons". John Wiley & Sons, USA.

Kelkar, R.R. and Yadav, B.S. "A basic guide for the use of INSAT imagery". IMD, New Delhi.

Moser, W. and Raschke, E. 1984. "Incident solar radiation over Europe estimated from METEOSAT data." *Journal of Climate and Applied Meteorology,* 23, 166-170.

Pinker, R.T. and Ewing, J.A. "Modeling surface solar radiation : Model formulation and validation". *Journal of Climate and Applied Meteorology,* 24, 389-401.

Price, J.C. 1984. "Land surface temperature measurements from the split window channels of the NOAA-7 AVHRR". *J. Geophysics Res.,* 89, 7231-7237.

Smith, W.L., Woolf, H.M., Hayden, C.M., Mark, D.Q. and Mc Milan, L.M. 1979. "The TIROS-N operational vertical sounder". *Bull. Am. Soc.,* 60, 1177-1187.

Tarpley, J.D. 1979. "Estimating incident solar radiation at the surface from geostationary satellite data". *J. Applied Meteorology,* 18, 1172-1181.

20

Monitoring Soil Erosion with GIS

Soil degradation by accelerated water and wind-induced erosion is a serious problem and will remain so during the 21st century, especially in developing countries of tropics and subtropics. Erosion is a natural geomorphic process occurring continually over the earth's surface. However, the acceleration of this process through anthropogenic perturbations can have severe impacts on soil and environmental quality. Accelerated soil erosion has adverse economic and environmental impacts. Economic effects are due to loss of farm income due to on-site and off-site reduction in income and other losses with adverse impact on crop/animal production. The on-site and off-site effects of soil erosion on productivity are depicted in Figure 1 and Figure 2, respectively. Off-site economic impact of soil erosion is presented in Figure 3. Table 1 shows regional food production statistics for 1995 with and without soil erosion in the world. The data in Table 1 indicate total loss of food production at 31 M Mg for Africa, 190 M Mg for Asia and 18 M Mg for tropical America.

Table 1. Regional food production statistics for 1995 with (a) and without (b) soil erosion.

Region	Cereals X 1000000		Soybeans X 1000000		Pulses X 1000000		Roots and tubers X 1000000	
	A	B	A	B	A	B	A	B
North Central America	358	376(5)	61	64(5)	6	65(b)	28	29(5)
Europe	268	281(5)	1	1(5)	6	6(5)	80	84(5)
Oceania	27	28(10)	—	—	2	2(15)	3	3(10)
Africa	100	110(10)	0.5	0.6(20)	7	8(20)	135	155(15)
Asia	929	1068(15)	21	23(10)	27	31(15)	248	293(18)
South America	90	99(10)	41	45(10)	4	4(10)	46	51(12)
Others	124	130(5)	3	3(5)	4	4(5)	69	72(5)
Total	1896	2092	126	136	50	61	609	687

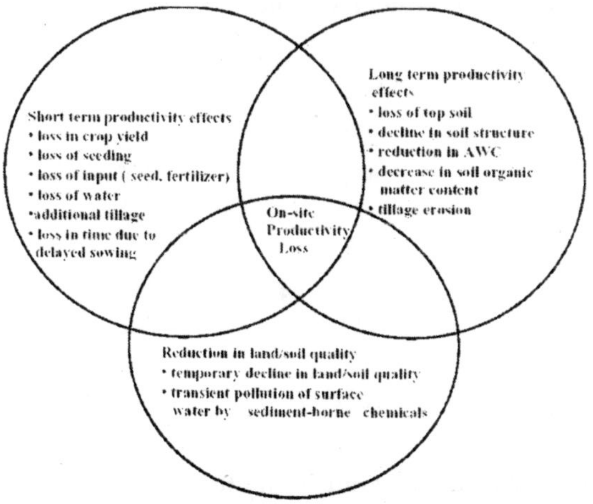

Figure 1: On-site effects of soil erosion on productivity are due to short-term and long-term effects, and on decline in soil quality.

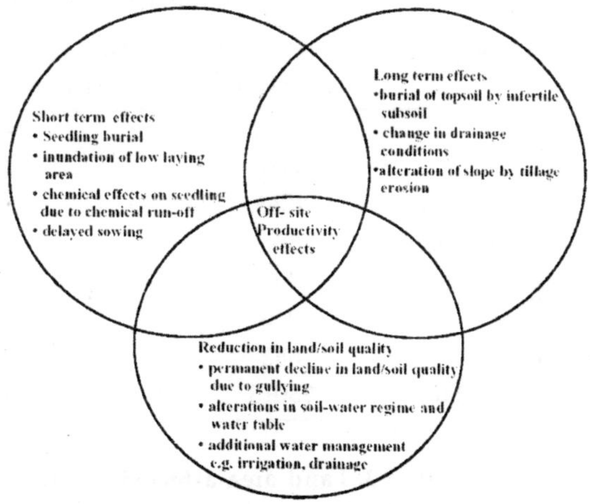

Figure 2: Off-site effects of soil erosion on productivity may be due to short-term or long-term and due to decline in land/soil quality.

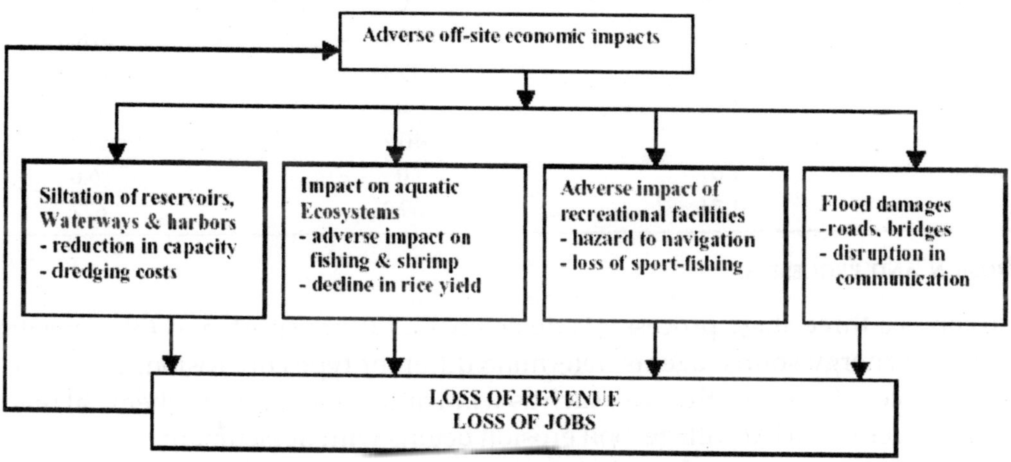

Figure 3: Off-site economic impact of soil erosion

GLOBAL EXTENT OF SOIL DEGRADATION BY EROSION

The total land area subjected to human-induced soil degradation is estimated at about 2 billion ha. Of this, the land area affected by soil degradation due to erosion is estimated at 1100 Mha by water erosion and 550 Mha by wind erosion (Table 2). South Asia is one of the regions in the world where soil erosion by water and wind is a severe problem (Table 3).

Table 2. Global extent of human–induced soil degradation.

World Regions	Total Land Area (10 ha)	Human induced soil degradation (10 ha)	Soil erosion (10 ha)	
			Water	Wind
Africa	2966	494	227	186
Asia	4256	748	441	222
South America	1768	243	123	42
Central America	306	63	46	5
North America	1885	95	60	35
Europe	950	219	114	42
Oceania	882	103	83	16
World Total	13013	1965	1094	548

Table 3. Land area affected by soil erosion by water and wind in South Asia.

Country	Water erosion (Mha)	Wind erosion (Mha)	Total land area (Mha)
Afghanistan	11.2	2.1	65.3
Bangladesh	1.5	0	14.4
Bhutan	0.04	0	4.7
India	32.8	10.8	328.8
Iran	26.4	35.4	165.3
Nepal	1.6	0	14.7
Pakistan	7.2	10.7	79.6
Sri Lanka	1.0	0	6.6
Total	81.74	59.0	677.

SOIL EROSION AND PROCESSES

Soil erosion is a three stage process: (1) detachment, (2) transport, and (3) deposition of soil. Different energy source agents determine different types of erosion. There are four principal sources of energy: physical, such as wind and water, gravity, chemical reactions and anthropogenic, such as tillage. Soil erosion begins with detachment, which is caused by break down of aggregates by raindrop impact, sheering or drag force of water and wind. Detached particles are transported by flowing water (over-land flow and inter-flow) and wind, and deposited when the velocity of water or wind decreases by the effect of slope or ground cover. Three processes viz. dispersion, compaction and crusting, accelerate the natural rate of soil erosion. These processes decrease structural stability, reduce soil strength, exacerbate erodibility and accentuate susceptibility to transport by overland flow, interflow, wind or gravity. These processes are accentuated by soil disturbance (by tillage, vehicular traffic), lack of ground cover and harsh climate.

FACTORS OF SOIL EROSION

The soil erosion process is modified by biophysical environment comprising soil, climate, terrain and ground cover and interactions between them (Figure 4). Soil erodibility –

susceptibility of soil to agent of erosion - is determined by inherent soil properties e.g., texture, structure, soil organic matter content, clay minerals, exchangeable cations and water retention and transmission properties. Climatic erosivity includes drop size distribution and intensity of rain, amount and frequency of rainfall, run-off amount and velocity, and wind velocity. Important terrain characteristics for studying soil erosion are slope gradient, length, aspect and shape. Ground cover exerts a strong moderating impact on dissipating the energy supplied by agents of soil erosion. The effect of biophysical processes governing soil erosion is influenced by economic, social and political causes (Figure 4).

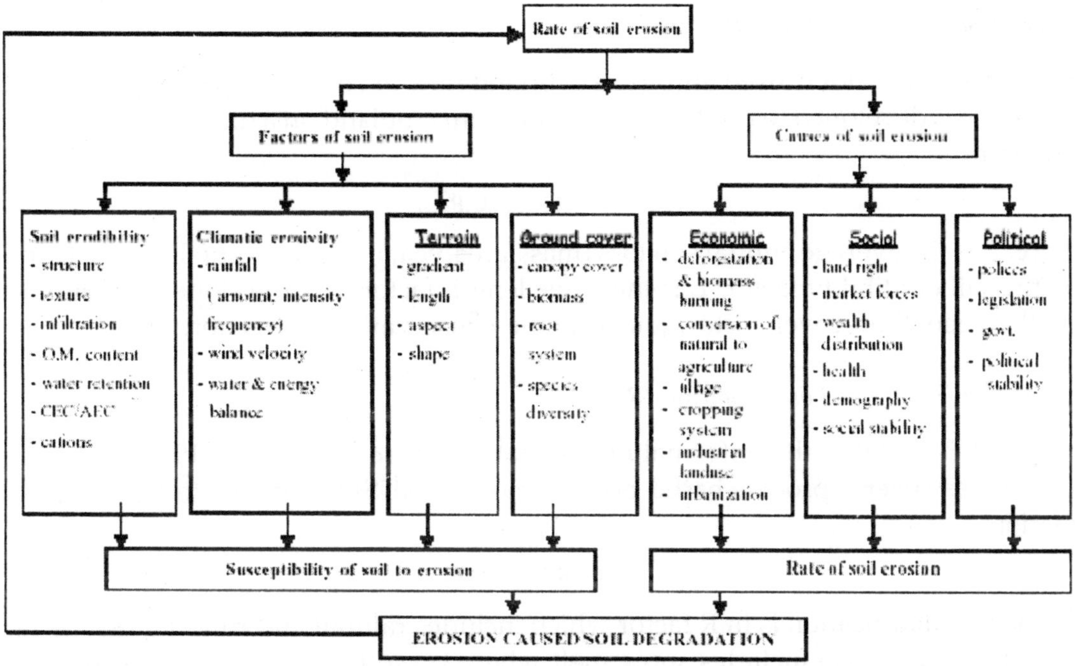

Figure 4: Factors of soil erosion; causes of soil erosion and interactions between them.

MODELLING SOIL EROSION

Field studies for prediction and assessment of soil erosion are expensive, time-consuming and need to be collected over many years. Though providing detailed understanding of the erosion processes, field studies have limitations because of complexity of interactions and the difficulty of generalising from the results. Soil erosion models can simulate erosion processes in the watershed and may be able to take into account many of the complex interactions that affect rates of erosion. Soil erosion prediction and assessment has been a challenge to researchers since the 1930s' and several models have been developed. These models are categorised as empirical, semi-empirical and physical process-based models. Empirical models are primarily based on observation and are usually statistical in nature. Semi-empirical model lies somewhere between physically process-based models and empirical models and are based on spatially lumped forms of water and sediment continuity equations. Physical process-based models are intended to represent the essential mechanism controlling erosion. They represent the synthesis of the individual components which affect erosion, including the complex interactions between various factors and their spatial and temporal variabilities. Some of the widely used erosion models are discussed below:

Empirical Models

Universal Soil Loss Equation (USLE)

USLE is the most widely used empirical overland flow or sheet-rill erosion equation. The equation was developed to predict soil erosion from cropland on a hillslope. The equation is given by –

$$A = R.K.L.S.C.P$$

Where, A is the average annual soil loss (mass/area/year); R is the rainfall erosivity index; K is the soil erodibility factor; L is the slope length factor; S is the slope gradient factor; C is the vegetation cover factor, and P is the conservation protection factor.

Revised Universal Soil Loss Equation (RUSLE)

The RUSLE updates the information on data required after the 1978 release, and incorporates several process-based erosion models. RUSLE remains to be a regression equation –

$$A = R.K.L.S.C.P$$

A principal modification is in R factor which includes rainfall and runoff erosivity factor (run-off erosivity also includes snow melt where run-off is significant). There are also changes in C factor which is based on computation of sub-factor called soil loss ratios (SLR). The SLR depends on sub-factors : prior landuse, canopy cover, surface cover, surface roughness and soil moisture.

Semi-empirical Models

Modified Universal Soil Loss Equation (MUSLE)

Williams proposed a modified version of USLE that can be written as–

$$Sye = Xe.K.L.S.Ce.Pe$$

Where, Sye is the event sediment yield

$$Xe = \alpha.\ (Qe.\ qp)\ 0.56$$

Where, α is an empirical co-efficient; Qe is the run-off amount and qp is the peak run-off rate obtained during the erosion event and K.L.S.Ce & Pe are as defined for USLE.

Morgan, Morgan and Finney (MMF) Model

Morgan *et al.* developed a model to predict annual soil loss which endeavors to retain the simplicity of USLE and encompasses some of the recent advances in understanding of erosion process into a water phase and sediment phase. Sediment phase considers soil erosion to result from the detachment of soil particles by overland flow. Thus, the sediment phase comprises two predictive equations, one for rate of splash detachment and one for the transport capacity of overland flow. The model uses six operating equations for which 15 input parameters are required (Table 4). The model compares predictions of detachment by rain splash and the transport capacity of the run-off and assesses the lower of the two values as the annual rate of soil loss, thereby denoting whether detachment or transport is the limiting factor.

Table 4. Operative functions and input parameters of Morgan, Morgan & Finney Soil erosion model

Operative functions		Parameter		Description
Water phase:		*E*	—	*kinetic energy of rainfall (J/m^2)*
$E = R*(11.9 + 8.7*\text{Log } I)$	(1)	*Q*	—	*volume of overland flow (mm)*
$Q = R*\exp(-R_C/R_0)$	(2)	*F*	—	*rate of detachment by raindrop impact (kg/m^2)*
Where,		*G*	—	*transport capacity of overland flow (kg/m^2)*
$R_C = 1000 * MS * BD * RD * (E_t/E_0)^{0.5}$	(3)	*R*	—	*Annual rainfall (mm)*
$R_0 = R/R_n$	(4)	*Rn*	—	*Number of rainy days in the year*
Sediment phase:		*I*	—	*Intensity of erosive rain (mm/h).*
$F = K*(E*e^{-0.05*A})*10^{-3}$	(5)	*A*	—	*Percentage of rainfall contributing to permanent interception and stream flow (%).*
$G = C * Q^2 * \sin S * 10^{-3}$	(6)	E_t/E_0	—	*Ratio of actual (E_t) to potential (E_0) evaporation.*
		MS	—	*Soil moisture content at field capacity or 1/3 bar tension (% w/w).*
		BD	—	*Bulk density of the top layer (Mg/m^3)*
		RD	—	*Top soil rooting depth (m) defined as the depth of soil from the surface to an impermeable or stony layer, to the base of A horizon; to the dominant root base*
		K	—	*Soil detachability index (g/J) defined as the weight of soil detached from soil mass per unit of rainfall energy.*
		S	—	*Steepness of the ground slope expressed as slope angle.*
		C	—	*Crop cover management factor. Combines C and P factors of the USLE*

Physical Process-based Model

Empirical models have constraints of applicability limited to ecological conditions similar to those from which data were used in their development. Further, USLE cannot deal with deposition; its applicability limits large areas and watersheds. Based on these considerations, several process-based models have been developed (e.g. WEPP, EUROSEM, LISEM).

Water Erosion Prediction Project (WEPP) Model

WEPP is an example of widely used physically process-based erosion model. It was developed as a system modeling approach for predicting and estimating soil loss and selecting catchment management practices for soil conservation. Basic erosion and deposition equations in WEPP are based on the mass balance formulation that uses rill and inter-rill concept of soil erosion, which is a steady-state sediment continuity equation. The WEPP model computes erosion by rill and inter-rill processes. The sediment delivery to rill from inter-rill is computed by following equation–

$$Di = Ki.\ Ie^2.\ Ge.\ Ce.\ Sf$$

Where, Di is the delivery of sediment from inter-rill areas to rill (kg/m²/ sec); Ki is the inter-rill erodibility (kg/m⁴/sec); Ie is the effective rainfall intensity (m/sec.); Ge is the ground cover adjustment factor and Sf is the slope adjustment factor calculated as per equation given below –

$$Sf = 1.05 - 0.85 \exp(-4 \sin \alpha)$$

Where, α is the slope of the surface towards nearby rill. In comparison, rill erosion is the detachment and transport of soil particles by concentrated flowing water –

$$Dc = Kr.\ (T - Tc)$$

Where, Kr is the rill erodibility (sec/m); T is the hydraulic shear of flowing water (Pa) and Tc is the critical hydraulic shear that must be exceeded before rill detachment can occur (Pa).

Wind Erosion Model

Comparable to the USLE, a wind erosion model was proposed by Woodruff and Siddoway (1965) as shown in equation given below –

$$E = f(I, K, C, L, V)$$

Where, E is the mean annual wind erosion; I is the soil erodibility index; C is the climatic factor (Wind energy); L is the unsheltered median travel distance of wind across a field; V is the equivalent vegetative cover. This equation has been widely adopted and used for estimating erosion hazard in dry lands.

USE OF SATELLITE REMOTE SENSING AND GIS IN SOIL EROSION MODELING

The potential utility of remotely sensed data in the form of aerial photographs and satellite sensors data has been well recognised in mapping and assessing landscape attributes controlling soil erosion, such as physiography, soils, land use/land cover, relief, soil erosion pattern. Remote Sensing can facilitate studying the factors enhancing the process, such as soil type, slope gradient, drainage, geology and land cover. Multi-temporal satellite images provide valuable information related to seasonal land use dynamics. Satellite data can be used for studying erosional features, such as gullies, rainfall interception by vegetation and vegetation cover factor. DEM (Digital Elevation Model) one of the vital inputs required for soil erosion modeling can be created by analysis of stereoscopic optical and microwave (SAR) remote sensing data. Geographic Information System (GIS) has emerged as a powerful tool for handling spatial and non-spatial geo-referenced data for preparation and visualisation of input and output, and for interaction with models. There is considerable potential for the use of GIS technology as an aid to the soil erosion inventory with reference to soil erosion modeling and erosion risk assessment.

Erosional soil loss is most frequently assessed by USLE. Spanner et al. first demonstrated the potential of GIS for erosional soil loss assessment using USLE. Several studies showed the potential utility of RS and GIS techniques for quantitatively assessing erosional soil loss. Satellite data analyzed soil and land cover maps and DEM derived and ancillary soil and agro-climatic rainfall data are the basic inputs used in USLE for computation of soil loss. Kudrat and Saha showed the feasibility of GIS to estimate actual and potential sediment yields following Sediment Yield Prediction Equation (SYPE) using RS derived soil and land use information, DEM derived slope and ancillary rainfall and temperature data. MMF model was used for quantification of soil loss by water erosion in Doon Valley, Dehra Dun, India, in GIS environment using various satellite remote sensing derived inputs. The availability of GIS tools and more powerful computing facilities makes it possible to overcome difficulties and limitations and to develop distributed 326 Water and Wind induced Soil Erosion Assessment and Monitoring continuous time models, based on available regional information. Recent development of deterministic models provides some spatially distributed tools, such as AGNPS; ANSWERS, and SWAT. The primary layers required for soil erosion modeling are terrain slope gradient and slope length which can be generated by GIS aided processing of DEM. Flanagan *et al.* generated the necessary topographic inputs for soil erosion and model simulations by linking WEPP model and GIS and utilising DEM.

REFERENCES

ASD 2001. Agriculture & Soils Division, IIRS P.G. Diploma Course pilot project report on Land Evaluation for Landuse Planning by integrated use of Remote Sensing and GIS –a case study of Bhogabati Watershed in Kolhapur district.

Lal, R. 1998. "Soil erosion impact on agronomic productivity and environment quality: Critical Review". *Plant Science,* 17 : 319 – 464.

Pande, L.M., Prasad, J., Saha, S.K. and Subramanyam, C. 1992. "Review of Remote Sensing applications to soils and agriculture". *Proc. Silver Jubilee Seminar,* IIRS, Dehra Dun.

21

GIS and Agrometeorological Services

Remote sensing techniques have been operationally used in many countries to provide basic information on crops, soils, water resources and the impact of drought and flood on agriculture. Integrated studies on soil and water conservation using remote sensing and GIS have been progressing with a view to raising agricultural production. In India procedures for pre-harvest acreage estimation of major crops such as wheat, rice and sorghum, using sampling and digital techniques based on remotely sensed data have improved greatly. The question that arises is in what respect can remote sensing be used in agricultural information systems. In fact agricultural production encompasses myriad activities related to crops, soils, water resources, climate and local weather of the region. For some of the specific themes of these disciplines, remote sensing technology has been observed to be operational and in some others some more research may be needed to enable integration into the total agricultural information system. Mainly, an agricultural resources inventory may need some knowledge of geology and geomorphology, forestry and vegetation, land use, and land degradation information, agricultural crops, water resources and agrometeorology. Ever since satellite remotely sensed data was available in digital format, the use of computers for analysis and interpretation of the data took roots. As the data actually depicted the state of the land as it existed at the time of observation, it was like bringing the ground to the office in the form of a picture to be studied by the various specialists. Data sets available from a variety of satellites have opened up tremendous possibilities of extracting a variety of information. In several cases of resources studies, the technique is operational, catering to several varied needs in the area of management of natural resources. The major advantage in using satellite remotely sensed data is that it is the latest available at any given time.

AGRO-METEOROLOGICAL INFORMATION BASED ON REMOTE SENSING DATA

Application potentials of remote sensing techniques on resources related to some important agrometeorological services are mentioned here :

1. *Agriculture (crops):* One basic information that remote sensing can provide to

agriculture is data related to crop identification and area measurement under different types of crops, or acreage estimation. This enables us to somewhat estimate the total production by understanding the yield per unit area. Such information has far-reaching consequences in providing adequate food security.

2. *Forestry and vegetation mapping:* Remote sensing can aid in providing:
   (a) information about the extent of forest cover and give a general idea of the types of forest cover;
   (b) forest canopy density condition ;
   (c) detection of forest hazards like fire, disease and excessive felling.
3. *Water resources :* Understanding water resources is important from agricultural point of view. Water supply to agriculturally related sectors depends upon the available resources, both in terms of quantity and quality. Remote sensing data is useful in assessing water resources, irrigated area studies and its monitoring and determining potential ground water zones.

Remote sensing can also provide data related to ocean and coastal zones like identifying potential area of fish concentration, environmental degradation that takes place in coastal zones due to over exploitation, etc. Other promising areas of applications include, disaster assessment, drought monitoring, environmental monitoring, etc. all of which have advanced significantly.

ASSESSMENT OF CROP CONDITION AND ESTIMATION OF YIELDS BY REMOTE SENSING TECHNIQUE

The use of remote sensing data for estimating crop acreage estimation has reached a stage near operational level. Studies carried out for estimating acreage under different crops in many countries show a near 90 percent accuracy level. In many countries, production forecasting of certain crops, crop yield modeling and crop stress detection are done using remote sensing data. Yield is influenced by many factors, such as crop genotype, soil characteristics, cultural practices adopted, meteorological conditions and influences of diseases and pests. Many approaches have been followed to determine the integrated effects of various parameters that affect crop growth and crop yield. Several yield models have been developed in which data obtained from various types of satellites to cover some of the parameters have been used. A major constraint however is a cloudy sky during the cropping season when normal optical remote sensing cannot give good data. However with the emergence of microwave remote sensed data, this can be overcome as such instruments can penetrate through clouds. To ensure complete success in predicting yields, some more research experiments may be needed to evolve a foolproof system.

Many countries have developed methods to assess crop growth and development from several sources of information such as, surveys of farm operators, crop condition reports from field surveys and local weather information. Remote sensing technology can provide supplemental spatial data to provide timely information on crop condition and potential yields. The timely evaluation of potential yields is increasingly important because of the growing economic impact of agricultural production on world markets. The use of the NDVI parameter to estimate crop yields is a specific extension of the above general concept. The seasonal accumulated NDVI values correlate well with the reported crop yields in semi-arid regions. Crop growth simulation models have been successfully used for predicting crop yields at the field level. However, numerous input requirements that are specific to the crop type, soil characteristics and management practices limit their applicability for regional studies. Integrating parameters derived from remotely sensed data with a growth model provides spatial integrity and near real time "calibration" of crop growth simulations. Remotely sensed data are incorporated in simulations of agricultural crop yields to calibrate or adjust model parameters during the simulation period to ensure agreement between the modeled and satellite observed parameters.

REAL TIME FORESTRY INFORMATION FROM SATELLITE DATA

The use of satellite remote sensing data has been found to be a reliable and useful tool for gathering forest information. Gathering of real time forestry information enables us to compile gross forest vegetation resources data in a single format and monitor the changes in the areal coverage of the forest. Since remote sensing satellites re-observe the same areas at periodic intervals, we have the added advantage of monitoring changes that occur in the area of coverage, to plan for remedial measures for any adverse happening, much sooner than what we were able to do in the past, when the facility of remote sensing from space altitudes was not available. It is in this context that we could view forest cover or vegetation cover of a large area, and the changes that occur to it, as the forest is a renewable natural resource which is very vital for human survival. The type of exercise also gives us the approximate rate at which the forest cover is changing and focusses attention on what should be done to ensure sustainability.

IDENTIFICATION OF FISHING ZONE BY REMOTE SENSING

Remote sensing methods have helped greatly in the optimisation of ocean resources. As 60 to 70 per cent of the world's population live within 20-30 km of the coastline, coastal zone management and optimisation of ocean resources have grown in importance. With the advent of remote sensing methods using satellite and aerial survey, the data coverage and accessibility have increased. Several parameters relating to the oceans including fisheries can be studied using satellite and aircraft data. While other ocean related applications have become possible on remotely sensed data from satellites, on a near real-time basis, here, it has been shown how a particular satellite and the sensor mounted on it have been used to obtain sea surface temperature and map it, on a regular basis, and pass it on to the fishermen who could concentrate on high potential areas and improve the catch.

The ocean colour as measured by the sensor is found to bear a direct relationship with the suspended material of the water, thermal characteristics and the location of greenish biological matter. One of the important parameters that can be measured with sufficient accuracy is the Sea Surface Temperature (SST), which has related to the concentration of fish population. This is an example of how a high technology has been applied at the grass-root level increasing the earning capacity by increased fish catches.

FISHING ZONE MAPS

SST derived from NOAA-AVHRR satellite serves as a very useful indicator of prevailing and changing environmental conditions and is one of the important parameters which decides suitable environmental conditions for fish aggregation. SST images obtained from satellite imagery over three or four days are composited and the minimum and maximum temperatures are noted down. These values are processed to obtain maximum contrast of the thermal information. This information is used to prepare relative thermal gradient images.

From these images, features such as thermal boundaries, relative temperature gradients to a level of 1 degree centigrade, level contour zones, eddies and upwelling zones are identified. These features are transferred using optical instruments to corresponding sectors of the coastal maps prepared with the help of Naval Hydrograph charts. Later, the location of the Potential Fishing Zone (PFZ) with reference to a particular fishing centre is drawn by identifying the nearest point of the thermal feature to that fishing centre. The information extracted consists of distance in kilometres, depth (for position fixing) in metres and bearing in degrees with reference to the North for a particular fishing centre.

Monitoring Natural Disasters by Remote Sensing

While aerial remotely sensed data were used for a long time, it is the satellite based remote sensing data which, because of its continuous availability and capacity to observe large areas, is considered a powerful medium to monitor the changes in the Earth's environment and take timely action. Remote sensing satellite information helps minimise damages e.g. the death of cattle, humans etc. and the damage of agricultural production in time of natural calamities by early warning system. Floods occur mainly due to heavy rainfall in association with low pressure, depressions and cyclones. While floods and cyclones cannot be totally eliminated, careful monitoring and planning can certainly mitigate the destruction and help in evolving suitable rehabilitation measures based on remotely sensed data. Attempts are being made to evolve a drought prediction system using remotely sensed data, but drought prediction is difficult, and a foolproof mechanism will probably take time. However, the severity of droughts can easily be assessed, thus providing information to the authorities for implementing relief measures.

Forest fires are considered a potential hazard with physical, biological, ecological and environmental consequences. Forest fires occur frequently in tropical countries particularly in the dry and hot seasons causing serious damage to the forest resources and agricultural production. Since the number of forest fires are increasing every year, continuous monitoring is of great importance, not only to understand present trends but also to devise a model to predict the possibility of fires in future. In a recent fire in the Rajiv Gandhi National Park situated in South Karnataka (India), remote sensing data was used for studying environmental aspects and the results were very encouraging.

MONITORING PESTS AND DISEASE FROM SATELLITE DATA

Various factors such as intensive cultivation, monocropping, changing weather conditions and indiscriminate use of pesticides have resulted in frequent outbreaks of crop pests and diseases causing huge crop losses. Minimising these losses is one way of enhancing grain production and remote sensing tool has been found very useful in monitoring large areas frequently. The Earth observing systems are useful in monitoring weather and ecological conditions favourable for crop pests and diseases. Weather conditions such as temperature, humidity (moisture), sunshine hours (light) and wind play major influence on the densities of pest population and their natural enemies. Among the weather parameters that can be remotely sensed, type of cloud, extent of cloud cover, cold cloud duration (a surrogate for rainfall) are the most easily retrievable. Such information was used by phytopathologists to study rust diseases of wheat crop. An aircraft fitted with a camera loaded with colour infrared films and flown over Kerala state (India), identified coconut areas severely affected by 'wilt' which could not be easily detected from the ground. This gave a clue to the area that could be viewed from satellite altitude.

Understanding the magnitude of crop losses is necessary to appreciate the importance of plant protection in crop production programmes. Losses can be due to biotic factors such as pests/diseases/weeds and abiotic factors such as drought, floods, cyclones and hailstorms. Damage caused by pests may be quantitative (overall reduction in yield), or qualitative such as change in colour and offensive odour. The regional disparities in crop condition assessment, the complex Centre-State relationships in handling relief measures and the introduction of crop insurance scheme, call for an unbiased, objective and timely information system to give early warning, to indicate the intensity of such hazards and to assess the loss. FAO had organised an international training programme at National Remote Sensing Agency (NRSA) in India a decade ago, involving countries like India, Pakistan and Tanzania in the use of various remote sensing data to identify locust-breeding areas. The directorate of plant protection handles such issues. There is no proper information about the area affected by pests and diseases and other yield reducing factors on an all-India basis. It is stated that the brown plant hopper (BPH) of rice is one of the dreaded insect pests in Asia. BPH is stated to be associated with synoptic weather conditions (depressions). Double cropping, extensive rice cultivation in the command area and indiscriminate use of fertilisers and pesticides aid the occurrence of BPH.

DESERT LOCUST FORECASTING

Desert locust plagues affect about 20 percent of the earth's surface spreading across Africa, the middle east and south-west Asia. They breed in areas that have sufficient soil moisture and vegetation to support the early stages of this insect (viz. egg laying and hopper development). They migrate from west to east along with the passage of troughs moving in the westerlies and northward and southward along with the Inter-tropical Convergence Zone (ITCZ). The main weather systems bringing rainfall favourable for the development of desert locust are western disturbances, depressions over Arabian Sea and a few depressions developed over land.

Remotely sensed vegetation indices and rainfall estimates based on cold cloud duration and other cloud indexing techniques are the only cost-effective methods to survey the vast stretches of desert locust habitat. A few studies have focussed on the collection of historical data on weather and habitat conditions with the dynamics of locust development stages, and synthesis of the data using Geographical Information System (GIS) and evolving decision support systems. This system integrates remotely sensed landform soil texture, soil moisture and vegetation density with the daily weather data to forecast the suitable breeding sites and time of onset of locust upsurge in and around the study area.

REMOTE SENSING AND DROUGHT MONITORING

One of the natural calamities that affects us is lack of normal rainfall and consequent drought conditions which in turn affect agricultural productivity. Drought conditions can be monitored using data obtained from satellite. This system provides efficient and timely monitoring capability by integrating the timeliness and objectivity of space observations with details of ground perceptions. During drought, physiognomic changes of vegetation may become apparent. Satellite sensors are capable of discerning many such changes through spectral radiance measures and manipulation of such measures into vegetation indices, which are sensitive to the rate of plant growth as well as to the amount of growth. Such indices are also sensitive to the changes in vegetation affected by moisture stress. The visible and near infrared bands on the satellite multispectral sensors allow monitoring of the greenness of vegetation. This property is used in the case of monitoring drought, as the stressed vegetation and other bare ground, water, etc. reflect differently. Besides, moisture stress in vegetation, resulting from prolonged rainfall deficiency, is reflected by lowering of vegetation index values. Such decrease could also be caused by other stresses such as pest/disease attack, nutrient deficiency or soil geo-chemical effects. But this does not show up well in coarse resolution data which covers very large areas at a time.

Reliable drought interpretation requires a Geographical Information System (GIS) based approach, since the topography, soil type, spatial rainfall variability, crop type and variety, irrigation support and management practices are all relevant parameters. In recent years, many investigations have demonstrated the capability of satellite-borne sensors to provide information on various drought indicators, which helps to monitor drought more effectively. The following paragraphs discuss remote sensing of rainfall, soil moisture, and vegetation/crop conditions, which are helpful in delineating agricultural drought.

RAINFALL ESTIMATION BY REMOTE SENSING

Satellite estimation of rainfall is not likely to be better than rainfall measured through conventional rain gauges, but nevertheless is useful to fill in spatial and temporal gaps in ground reports. Nageswara Rao and Rao demonstrated an approach for preparing an indicative drought map based on NOAA AVHRR derived rainfall estimation at the seedling stage of crop growth. For drought monitoring, quantitative point-specific rainfall estimates on the daily basis all over the country may not be required. What is needed, however, is a capability to spatially distribute the point rainfall observations over the areal unit in a qualitative manner.

REMOTE SENSING OF SOIL MOISTURE

Microwave sensors are probably the best soil moisture sensors, considering the strong physical relationship between the microwave response and soil moisture and the capability of microwaves to penetrate clouds, precipitation, and herbaceous vegetation. The principle advantage of active microwave sensors is that high spatial resolution can be obtained even at satellite altitudes. Microwave sensors can provide estimates of soil moisture only in surface layers up to 10 cm thick. This depth is too shallow, compared to the 1-2 m root zone of many field crops in the tropics. Using the water content in the top 10 cm of the surface layer, the moisture content can be calculated within acceptable limits and with minimum error when the surface soil moisture estimation is made just before dawn. Some investigations are under way at the National Remote Sensing Agency, Space Application Centre, and elsewhere to evaluate ERS- ISAR data for soil moisture estimates in the surface layers.

REMOTE SENSING OF VEGETATION STATUS

During periods of drought conditions, physiological changes within vegetation may become apparent. Satellite sensors are capable of discerning many such changes through spectral radiance measurements and manipulation of this information into vegetation indices, which are sensitive to the rate of plant growth as well as to the amount of growth. Such indices are also sensitive to the changes in vegetation affected by moisture stress. The visible and near infrared (IR) bands on the satellite multispectral sensors allow monitoring of the greenness of vegetation. Stressed vegetation is less reflective in the near IR channel than nonstressed vegetation and also absorbs less energy in the visible band. Thus the discrimination between moisture stressed and normal crops in these wavelengths is more suitable for monitoring the impact of drought on vegetation. The NDVI varies with the magnitude of green foliage (green leaf area index, green biomass, or percentage green foliage ground cover) brought about phenological changes or environmental stresses. The temporal pattern of NDVI is useful in diagnosing vegetation conditions Moisture stress in vegetation, resulting from prolonged rainfall deficiency is reflected by lower NDVI values. Such a decrease could also be caused by other stresses, such as pest/disease infestation, nutrient deficiency, or soil geochemical effects. Discrimination of moisture stress from other effects does not present a problem in coarse resolution data over large areal units, as neither pest/disease attack nor nutrient stress is selected in terms of area or crop type.

NATIONAL AGRICULTURAL DROUGHT ASSESSMENT AND MONITORING SYSTEM (NADAMS)

Since 1989, National Agricultural Drought Assessment and Monitoring System (NADAMS) in India has been providing biweekly drought bulletins through the kharif season (June to December) for 246 districts in most of the peninsular and northern India. The bulletins, which describe prevalence, relative severity level, and persistence of drought through the season at the district level, are being sent to the concerned state and central administrators as well as to district-level officers. The drought assessment is based on a comparative evaluation of satellite observed green vegetation cover (both area and greenness) of a district in any specific time period to cover in similar periods in the previous year. The trend of seasonal vegetation development until the reporting period is also compared with trends of previous years. The drought interpretation takes into account rainfall and aridity anomaly trends. This nationwide early warning services has been found to be useful for providing early assessment of drought conditions.

DISSEMINATION OF THE INFORMATION THROUGH AGRO-MET ADVISORY SERVICE

Agrometeorological information is rarely provided as a finished product to the clients. Often it is used to complement the purely meteorological products, or delivered in combination with other remotely sensed products, such as information in soil wetness, land or vegetation cover (NDVI), likely presence of pests and/or diseases, estimates of the areal coverage of irrigated or flood-retreat crops, incidence of bush fires, etc. By the nature of their capacity to indicate the probable areal extent of a condition, and of the still very rapid evolution of the parameters that can be measured or derived, remotely-sensed data and their derived products will be a growing resource for the supply of agrometeorological products to clients. How information is delivered to the users of the product is, finally, of extreme importance. There are a number of issues that fall into information delivery. These include clearly defined users, user-friendly information, cooperation and coordination between producers and users of the product, proper training, and timeliness of information delivery. Hard copy publication sent via mail allows detailed text and graphics, but its effectiveness may be hampered by the timeliness of receipt. Delivery by radio allows rapid dissemination but limits the amount of information that may be provided. Internet technology combines the strengths of detailed information and rapid delivery but is definitely constrained by lack of Internet access in many developing countries. The product must be delivered or made accessible in sufficient time for the user to make professional sense from the information and use it appropriately in the management process. Formal lines of communication can be developed through user surveys and open forums. Information mechanisms, such as telephone, facsimile, or email exchange should also be encouraged.

Once established, such mechanisms for communication should become a routine occurrence to accommodate changing user needs, new technological innovations, and more efficient distribution procedures. Constructive feedback mechanisms promote an active dialogue to encourage improvements that not only technically enhance the bulletin but also increase its usefulness. Information delivery by Internet communication offers great opportunity to move quality products to the decision-maker rapidly. The computer age technology also allows efficient feedback mechanisms, which in turn may increase the demand for additional information.

REFERENCES

Das, H.P. 2000. "Monitoring the incidence of large scale droughts in India". pages 181-195. *In* Drought A Global Assessment(Donald A. Wilhite (Ed.)), Vol. 1, Routeldge London and New York.

Doraiswamy, P. C., Zara, P. and Stem, A. 1996. "Satellite remotely sensed data application in estimating crop condition and yields", pages 220 -240. *Remote Sensing Applications.* Narosa Publishing House.

Groten, S.M.E. 1993. "NDVI-Crop monitoring and early yield assessment of Burkina Faso". *International Journal of Remote Sensing,* 14(8): 1495-1515.

Gupta, R.K. 1993. "Comparative study of AVHRR ratio vegetation index and normalised difference vegetation index in district level agricultural monitoring". *International Journal of Remote Sensing* : 14(1): 55-73.

Healey, R.G., Robertson, S.G., Magor, J.I., Pender, J. and Gressman, K. 1996. "A GIS for Desert Locust Forecasting and Monitoring", *International Journal of Remote Sensing*: 10(1): 117-136.

Motha, R.P. 2001. "Agrometeorological Bulletins". How can we improve them. Proceedings of the Inter-Regional Workshop, Barbados.

Moulin, S., Fisher, A., Dedieu, G. and Delcolle, R. 1995. "Temporal variation in satellite reflectances at field and regional scales compared with values simulated by linking crop growth and SAIL models". *Remote Sensing of Environment,* 54: 261-272.

Nageswara Rao, P.P. and Rao, V.R. 1984. "An approach for agricultural drought monitoring using NOAA/ AVHRR and Landsat imagery". pages 225 –229. *In* Remote Sensing-From Research towards Operational Use (T.D. Guyenne and J.J. Hunt (eds)), (International Geoscience and Remote Sensing Symposium, Strasbourg, France, 27-30 August), Noordwijk, Netherlands : ESA Scientific and Technical Pulications Branch.

Bibliography

Ahuja, S.S., 2000, "Tea-production constraints stay". *In the Hindu-survey of Indian Agriculture,* 2000 pp. 105-108.

Arnold, J.G., Engel, B.A. and Srinivasan, R. 1998. "A continuous time grid cell watershed model". *Proc.* of application of Advanced Technology for management of Natural Resources.

ASD 2002. "Agriculture & Soils Division", IIRS P.G. Diploma Course pilot project report on Soil Erosion Inventory using IRS – WiFS data and GIS – a case study of Dehra Dun district (Uttaranchal).

Bacsi, Z., Thorton, P.K. and Dent, J.B. 1991. "Impacts of future climate change on Hungarian crop production: An application of crop growth simulation models". *Agric. Sys.,* 37: 435-450.

Badhwar, G. D., MacDonald, R.B., and Mehta, N.C. 1986. "Satellite-derived LAI and vegetation maps as input to global cycle models-a hierarchical approach". *Int. J. Remote Sensing, 7: 265-281.*

Beasley, D.B.C., Huggins, L.F. and Monke, E.J. 1990. "ANSWERS : a model for watershed planning". *Transactions of ASAE.* 23 : 938-944.

Bouman, B.A.M. 1995. "Crop modelling and remote sensing for yield prediction". *Netherland J. Agric. Sci.,* 43: 143-161.

Branden, Bagemen, J. and Agkiosobud, C.F., 1994, "Incentive-based non-point source pollution abatement in a re-authored clean water act". *Water Resource.* Japan Agricultural Software Association, 1996. Agriculture-related Software Book (In Japanese). Rakuyu Shobo, Tokyo.

Burrough, P.A. 1996. "Environmental modelling with geographical information system", pages 56-59. In: Models in Action, Quantitative Approaches in Systems Analysis (Stein, A., Penning de Vries, F.W.T. and Schotman, P.J. eds.), No 6, AB-DLO, Wageningen, The Netherlands.

Burrough, P.A. and McDonnell, R.A. 1998. "Principles of geographic information systems". Oxford University Press, Oxford, UK, pp. 10-16.

Calixte, J.P., Beinroth, F.J., Jones, J.W. and Lai, H. 1992. "Linking DSSAT to a geographic information system". *Agrotechnology Transfer,* 15: 1-7.

Carter, T.R. and Saariko, R.A. 1996. "Estimating regional crop potential in Finland under a changing climate". *Agric. For. Meteorol.,* 79: 301-313.

Chen, J.M. and Cihlar, J. 1996. "Retrieving leaf area index of boreal conifer forests using Landsat TM images". *Remote Sens.Environ.,* 55: 153-162.

Clevers, J.G.P.W., Buker, C., van Leeuwen, H.J.C. and Bouman, B.A.M. 1994. "A framework for monitoring crop growth by combining directional and spectral remote sensing information". *Remote Sens. Environ.,* 50: 161-170.

Dadhwal, V.K. 1999. "Remote Sensing and GIS for agricultural crop acreage and yield estimation". *Internat. Arch.* Photogramm. & Remote Sensing, XXXII, 7-W9, 58-67.

Dadhwal, V.K. and Ray, S.S. 2000. "Crop Assessment using remote sensing—Part II: Crop condition and yield assessment". *Indian J. Agric. Economics,* 55 (2, Suppl.), 55-67.

Delecolle, R. and Guerif, M. 1988. "Introducing spectral data into a plant process model for improving its prediction ability". Proceedings of the 4th International Colloquium Signatures Spectrales d'Objects en Teledetection, 18-22 Jan, 1988, Aussois, France, pp. 125-127.

Engel, T., Hoogenboom, G., Jones, J.W. and Wilkens, P.W. 1997. "AEGIS/WIN—a program for the application of crop simulation models across geographic areas". *Agronomy J.,* 89: 919-928.

Flanagan, D.C., Renschler, C.S. and Cochrane, T.A. 2000. "Application of the WEPP model with digital geographic information". 4th Int. Conf. on Integration of GIS and Environmental Modeling : Problems, Prospects and Research needs.

Gao, W. and Lesht, B.M. 1997. "Model inversion of satellite measured reflectances to obtain surface biophysical and bi-directional reflectance characteristics of grassland". *Remote Sens. Environ.,* 59: 461-471.

Guerif, M. and Duke, C. 1998. "Calibration of SUCROS emergence and early growth module for sugar beet using optical remote sensing data assimilation". *European J. Agronomy,* 9: 127-136.

Han, S., Evans, R.G., Hodges, T. and Rawlins, S.L. 1995. "Linking geographic information system with a potato simulation model for site-specific crop management". *J. Environ. Qual.,* 24: 772-777.

Hartkamp, D.A., White, J.W. and Hoogenboom, G. 1999. "Interfacing geographic information systems with agronomic modeling: A review". *Agron. J.,* 91: 761-772.

Horie, T., Yajima, M. and Nakagawa, H. 1992. "Yield forecasting". *Agric. For. Meteorol.,* 40: 211-236.

Iida, M., Umeda, M., Kaho, T., Lee, C.K. and Suguri, M., 1998, "Measurement of Annual Crops". International Conference on Precision Agriculture, St. Paul. MN. 19-22, July 1998, ASA, CSSA, and SSSA, Madison, WI.

Inoue, Y., Moran, M.S. and Horie, T. 1998. "Analysis of spectral measurements in paddy field for predicting rice growth and yield based on a simple crop simulation model". *Plant Production Sci.,* 1(4): 269-279.

Jones, J.W. 1993. ôDecision support systems for agricultural development, pages 459-471". In: *Systems Approaches for Agricultural Development,* Vol. 2. (Penning de Vries, F.W.T., Teng, P. and Metselaar, K. eds.), Kluwer Academic Publishers, Dordrecht, The Netherlands.

Knyazikhin, Y., Martonchik, J.V., Myneni, R.B., Diner, D.J. and Running, S.W. 1998. "Synergistic algorithm for estimating vegetation canopy leaf area index and fraction of absorbed photosynthetically active radiation from MODIS and MISR data". *J. Geophys. Res.,* 103: 32257-32275.

Lal, R. 1998. "Soil erosion impact on agronomic productivity and environment quality: Critical Review". *Plant Science,* 17: 319 – 464.

Maas, S.J. 1988a. "Use of remotely sensed information in agricultural crop growth models". *Ecol. Modelling,* 41: 247-268.

Maracchi, G., Perarnaud, V. and Kleschenko, A.D. 2000. "Applications of geographical information systems and remote sensing in agrometeorology". *Agric. Forest Meteorol.,* 103: 119-136.

Meyer-Roux, J. and Vossen, P. 1994. "The first phase of the MARS project, 1988-1993: overview, methods and results, pp. 33 – 85". In: Proceedings of the Conference on the MARS Project: Overview and Perspectives. Commission of the European Communities, Luxembourg.

Mongkolsawat, C., Thurangoon, P. and Sriwongsa 1994. "Soil erosion mapping with USLE and GIS". *Proc. Asian Conf. Rem. Sens.,* C-1-1 to C-1-6.

Morgan, R.P.C., Morgan, D.D.V. and Finney, H.J. 1984. "A predictive model for the assessment of erosion risk". *J. Agricultural Engineering Research* 30: 245 – 253.

Moulin, S. and Guerif, M. 1999. "Impacts of model parameter uncertainties on crop reflectance estimates: a regional case study on wheat". *Int. J. Remote Sens.,* 20(1): 213-218.

Myneni, R.B., Nemani, R.R. and Running, S.W. 1997. "Estimation of global leaf area index and absorbed PAR using radiative transfer models". *IEEE Trans. on Geosc. and Rem. Sensing,* 35(6): 1380-1393.

Nain, A.S., Dadhwal, V.K. and Singh, T.P. 2004. "Use of CERES-Wheat model for wheat yield forecast in central Indo-Gangetic plains of India". *J. Agricultural Science (camb.),* 142: 59-70.

Okano, C., Nishimune, A., Fukuhara, M., Okamoto, K. and Hayasaka, M., 1997, "Sugar beet nutritional diagnosis using remote sensing". P. 381-382. In T. Ando *et al.,* (ed) Plant nutrition for sustainable food production and environment. Kluwer Acad. Publi., Tokyo.

Pande, L.M., Prasad, J., Saha, S.K. and Subramanyam, C. 1992. "Review of Remote Sensing applications to soils and agriculture". *Proc. Silver Jubilee Seminar,* IIRS, Dehra Dun.

Pandya, M.R., Chaudhari, K.N., Singh, R.P., Sehgal, V.K., Bairagi, G.D., Sharma, R. and Dadhwal, V.K. 2003. "Leaf area index retrieval using IRS LISS-III sensor data and validation of MODIS LAI product over Madhya Pradesh". *Current Science,* 85(12): 1777-1782.

Pierce, J.F. and Peter Nowak, 1999, "Aspects of precision Agriculture". *Advances in Agronomy,* 67: 1-85.

Price, J.C. 1993. "Estimating leaf area index from satellite data. IEEE Trans". *Geoscience Remote Sensing,* 31: 727-734.

Qiu, J., Gao, W. and Lesht, B.M. 1998. "Inverting optical reflectance to estimate surface properties of vegetation canopies". *Int. J. Remote Sens.,* 19: 641-656.

Rastogi, A., Kalra, N., Agarwal, P.K., Sharma, S.K., Harit, R.C., Navalgund, R.R. and Dadhwal, V.K. 2000. "Estimation of wheat leaf area index from satellite data using Price model". *International J. Remote Sensing,* 21(15): 2943- 2949.

Ray, S.S., Panigrahy, P. and Parihar, J.S., 2001, "Role of Remote Sensing for precision farming–with special Reference to Indian Situation Scientific Note SAC/RESA/ARG/AMD/SN/01/2001", *Space Applications Center (ISRO),* Ahmedabad, pp: 1-21.

Renard, K.G., Foster, G.R., Lane, I.J. and Laflen, J.M. 1996. "Soil loss estimation". *In* Soil Erosion, Conservation and Rehabilitation; Agassi, M. (ed.). Marcel Dekkar, New York, 169-202.

Richardson, A.J., Wiegand, C.L., Arkin, G.F., Nixon, P.R. and Gerbermann, A.H. 1982. "Remotely sensed spectral indicators of sorghum development and their use in growth modelling". *Agric. Meteorol.,* 26: 11-23.

Sadler, E.J., Busscher, W.J., Bauer, P.J. and Karlen, D.J., 1998, "Spatial scale requirements for precision farming: A case study in the southeastern USA". *Agronomy Journal,* 90: 191-197.

Saha, S.K. and Pande, L.M. 1993. "Integrated approach towards soil erosion inventory for environmental conservation using satellite and agro-meteorological data". *Asia-Pacific Rem. Sens. J.,* 5(2) : 21-28.

Singh, G., Babu, R., Narain, P., Bhushan, L.S. and Abrol, I.P. 1992. "Soil erosion rates in India". *J. Soil Water Conservation,* 47: 93 95.

Spanner, M.A., Strahler, A.H. and Estes, J.E. 1982. *Proc. Int. Symp. Rem. Sens. Environ.,* Michigan, USA.

Srinivasan, 2001, "Precision farming in Asia: Progress and prospects, Geospatial Analysis Center, Regional Science Institute, Hokkaido, Japan". Wilson and Scot 1982, Concepts of variable rate technology with considerations for fertilizer application. *Journal of Production Agriculture,* 7(2): 195-201.

Thornton, P.K., Bowen, W.T., Ravelo, A.C., Wilkens, P.W., Farmer, G., Brock, J. and Brink, J.E. 1997. "Estimating millet production for famine early warning: an application of crop simulation modelling using satellite and ground-based data in Burkina". *Faso. Agric.* For. Meteorol, 83: 95-112.

Tim, U.S. 1996. "Coupling vadose zone models with GIS: Emerging trends and potential bottlenecks". *J. Environ. Qual.,* 25: 535-544.

Venkateswarlu, J. 1994. "Managing extreme stresses in arid zone of Western Rajasthan, India". *In* Stressed Ecosystem and Sustainable Agriculture. Virmani, S.M.; Katyal, J.C.; Eswaran, H. and Abrol, I.P. (eds.). Oxford & IBH Publishing, New Delhi, p. 161-171.

Vossen, P. 1995. "Early assessment of national yields: the approach developed by the MARSSTAT project on behalf of European Commission, pages 327-347". *In:* Proceedings of the Seminar on Yield Forecasting, 24-27 Oct., 1995, Villerfranche-sur-Mer, France.

Wade, G., Mueller, R., Cook, P.W. and Doraiswamy, P.C. 1994. "AVHRR Map products for Crop Condition Assessment: A Geographic Information Systems Approach". Photogrammetric Engineering & Remote Sensing, 60(9): 1145-1150.

Wiegand, C.L., Richardsons, A.J. and Kanemasu, E.T. 1979. "Leaf area index estimates for wheat from LANDSAT and their implications for evapotranspiration and crop modeling". *Agronomy J.*, 71: 336-342.

Williams, R. 1975. "Sediment Yield Prediction with USLE using run-off energy factor". *In* Present and prospective technology for predicting sediment yields and sources. ARS-S-40, USDA, Washington D.C. : 244-252.

Woodruff, N.P. and Siddoway, F..H. 1965. "A wind erosion equation". *Soil Sci. Soc. Am. Proc.*, 29: 602-608.

Young, R.A., Onstad, C.A., Bosch, D.D. and Anderson, W.P. 1989. "AGNPS—Agriculture-Non-Point Source Pollution Model : A watershed analysis tool". *Conservation Res.* Report 35, USDA, ARS, Morris, MN, USA.

Zhang, L. 1994. "A comparison of the efficiency of the three models to estimate water yield changes after forced catchment conversion". M.Sc. (Forest) Thesis, University of Melbourne, Australia.

Index